# THE SCIENCE OF
# JAMES BOND

# THE SCIENCE OF

# JAMES BOND

## THE SUPER-VILLAINS, TECH, AND SPY-CRAFT BEHIND THE FILM AND FICTION

## MARK BRAKE

Skyhorse Publishing

Skyhorse Publishing books may be purchased in bulk at special discounts for sales promotion, corporate gifts, fund-raising, or educational purposes. Special editions can also be created to specifications. For details, contact the Special Sales Department, Skyhorse Publishing, 307 West 36th Street, 11th Floor, New York, NY 10018 or info@skyhorsepublishing.com.

Skyhorse® and Skyhorse Publishing® are registered trademarks of Skyhorse Publishing, Inc.®, a Delaware corporation.

Visit our website at www.skyhorsepublishing.com.

10 9 8 7 6 5 4 3 2 1

Library of Congress Cataloging-in-Publication Data is available on file.

Cover design by Peter Donahue
Cover illustration by gettyimages

Print ISBN: 978-1-5107-4379-3
Ebook ISBN: 978-1-5107-4380-9

Printed in the United States of America

*This book is dedicated to my Dad—"Brake, James Brake"—the biggest Bond fan of them all.*

# CONTENTS

# INTRODUCTION

# SPY-FI CULTURE WITH A LICENCE TO KILL

As long as Ian Fleming's kind of idea of it is basically there, it is something which lasts. [Bond] has to be real, but he has to be also a kind of fantasy person, that kind of comic strip quality about it has to be there. Times have changed and we expect different things now.

—Dame Judi Dench, *The South Bank Show* (2008)

I think ever since the movies first appeared on the screens in the '60s, they created a new kind of cinematic genre, and I think that has lasted for several decades. Often, you'll pick up the newspaper and they'll refer to a villain as being someone who's Bondian, or some extraordinary piece of architecture that looks like it could be on a Bond set.

—Barbara Broccoli, *The South Bank Show* (2008)

Sitting in the movies as a kid, what a thrill it was to hear that incredible James Bond theme played live. The three distinct motifs echoed around the cavernous old movie theater: the surf rock guitar intro, followed by the rolling strings that peak and fall, and finally the raunchy seven-note riff in blaring brass, which now feels like the very definition of a traditional action-movie soundtrack. Such a musical trifecta would become the de facto daddy of the Bond franchise's sonic style.

Then, the opening sequence. A mysterious set of circles shift across a dark screen until they're resolved into the view down a rifled barrel. Our point of view is that of would-be assassin, as we see Bond walking in profile. But we're far too slow for Bond, who dramatically turns face-on and fires, as our cinematic vision blurs due to a descending curtain of blood. To my young impressionable eyes, it was like a bad dream, or some obscure animation from eastern Europe. Next, it was time for the title sequence. Today, the images of scantily clad females on whom the titles are wavily projected seems dated and inappropriate. And yet in my bug-eyed boyhood this was an adult world full of the erotic and exotic, the mysterious and the downright dangerous.

Bond, James Bond—the only international secret agent with a shelf life of fifty-seven years—and counting. Spy fiction had grown as a genre of literature in the early twentieth century. Espionage had been key to the context and plot. Stories had revolved around the rivalries and intrigues between the major powers, which had established modern intelligence agencies to administrate their power and imperialism. Then, before and after WWII, spy fiction was given new impetus by the conflicting ideologies of fascism and communism. The Cold War was a peak, with the emergence of global criminal organizations, rogue states, and world terrorist networks, as espionage became a potent threat to Western democracies.

James Bond was something different. Bond author and naval intelligence officer Ian Fleming had once dismissed his own creation as "bang, bang, bang, kiss, kiss." And yet Bond is far more than that. What sets Bond apart from the rest of spy fiction? The super-villains and locations? The gadgets and the girls? The cars and casinos? The martini—shaken, not stirred?

In this book, we shall think of the James Bond franchise as spy-fi, a genre of fiction that fuses spy fiction with science fiction, taking on board sci-fi's obsessions with super-villains, the future, world domination or destruction, and trends in science and tech, often in the form of gadgets, inventions, and spy devices. After all, as Daniel Craig says in the 2012 documentary *Everything or Nothing: The Untold Story of 007*, Bond is rarely about the past, and almost always about the future.

One of the most obvious and superficial aspects of the Bond franchise that qualifies as sci-fi is its fixation on gadgetry. Starting with *From Russia with Love*, we witness Bond's pre-mission science and tech talks with Q, and 007's chance to play in the sci-fi sandbox. Inventions and gadgets such as the typewriter-sized Lektor Decoder in *From Russia with Love*; the homing beacons of *Goldfinger*; *Thunderball's* iconic use of the jetpack; the rocket-gunned autogyro in *You Only Live Twice*; the electromagnetic RPM controller ring that guarantees a slot-machine win in *Diamonds Are Forever*; the watch that doubles as a buzz saw in *Live and Let Die*; the Lotus Esprit that sports missiles, landmines, and torpedoes in *The Spy Who Loved Me*; are all evidence that Bond toys with the kind of tech you might find decorating a science fiction movie.

Spy-fi is a genre that revolves around the adventures of a key character working as a secret agent or spy. For the most part, such adventures center on either the intrigue of espionage between rival superpowers (during the Cold War period, this was the West against either the USSR or China) or else trying to prevent a singular enemy super-villain, some diabolical mastermind such as Blofeld of SPECTRE, from achieving some fiendish plot. The content of Bond stories, whether on the page or on the silver screen, usually involve themes and settings that have as much to do with the outright fantasy of sci-fi as they do with ordinary espionage. One merely has to think of the outer space plots of *You Only Live Twice*, where both US and USSR space modules are stolen by a mysterious rogue agency under Blofeld, or *Moonraker*, where Drax Industries tries to exterminate the whole human race to wipe the slate clean and reboot Planet Earth anew! If the theme is not the final frontier of space, then it's another under-explored boundary, such as the deep-sea world in *The Spy Who Loved Me*, where Stromberg seeks not to conquer space, but to vanquish instead the seven-tenths of the world unexplored beneath the oceans. The spy-fi of James Bond doesn't present its spy fiction as pure and simple espionage, in the way that the Bourne franchise does, for example. Rather, Bond stories represent a reality that glamorizes spy-craft through its focus on a near future of science and high-tech, through corporate agencies and criminal organizations with almost unlimited resources and sky-high-stakes adventures.

(Another context for the Bond films was the process that some American film historians have described as "genre upscaling." In response to the huge dip in moviegoing and the decline in creative output from the movie studios, moviemakers focused on making fewer but bigger movies with higher production values. Genre was perfect for the job, so genre movies were at the cutting edge of Hollywood's new-found fetish with blockbusters in the 1950s and 1960s. Consequently, genres that had previously been thought of as low-budget, such as thrillers and sci-fi, now benefited from lavishing "A"-feature production values on "B"-movie topics. And with Bond we have the merging of the two into a new sub-genre of spy-fi. The pleasure for moviegoers consisted of finding themselves immersed in a game in which they knew the pieces and the rules, and drew delight simply from the minimal variations and nuances by which Bond realizes his mission. The spy-fi of Bond became typical of the escape machine geared for the entertainment of the masses.)

As Barbara Broccoli says, James Bond has been hugely influential to spy fiction and film in general. And, of course, spy-fi isn't limited to Bond. It can also be found in *Mission: Impossible* and *The Man from U.N.C.L.E.*, in both their 1960s TV series and modern cinematic formats; in the adult animated sitcom *Archer* and *Marvel's Agents of S.H.I.E.L.D.;* and in famous spy-fi satire series of movies such as *Austin Powers* and *Kingsman*. But they all come after Bond.

The point of this book is not to take the Bond tales and movies scientifically literally, as most other "Science of" books make the serious mistake of doing. Bond is spy *fiction*, after all. Nor will you find in these pages an obsession with inventions, a kind of commodity fetishism about the guns, gadgets, and boy toys that are mentioned in the Bond movies. They are mere decoration. No, this book looks at the bigger picture. The larger-than-life scientific and cultural contexts, which act as world-shaking scenarios to the Bond stories.

So, buckle up as we begin a spy-fi journey through the cinematic reels of one of history's most successful movie franchises. For each film, we shall follow Bond's progress while also looking at the bigger picture of science and tech in each plot. From the exploration of space in *Dr. No*, *You Only Live Twice*, *Moonraker*, and *GoldenEye*, through the nuclear paranoia of *For Your Eyes Only* and *Octopussy*, to the brave new world order of spying in *Skyfall* and *Spectre*. Fewer actors have played Bond than humans who've walked on the moon. And yet Bond has endured. He survived the Cold War. Opponents other than Russians were replaced as the years washed over him and we continued to forgive his taste for luxury cars, expensive watches, and martinis (shaken, not stirred). Like us, Bond lives in a changing world. Despite sociopolitical and cultural changes, Bond shines on—a self-controlled lone wolf trying to save a chaotic world from itself. His self-sacrificing and thankless tradecraft prevents him from having a real life and real relationships. And, as nuclear missile threats are replaced by a series of subtler threats in a globalized and digital world, Bond abides.

# PART I: 1962–1977

# DR. NO (1962)

# BEWARE OF THE BLACK-GLOVED BOFFIN

| Dr. No | Year: 1962 | Bond: Sean Connery | Director: Terence Young |
|--------|-----------|---------------------|--------------------------|
| **Producer:** Albert R. Broccoli & H. Saltzman | **Screenplay:** R. Maibaum, J. Harwood & B. Mather | **Distributor:** United Artists | **Release Date:** October 5 |
| **Running Time:** 109 minutes | **Budget:** $1.1 million | **Box Office:** $59.5 million | **Body Count:** 12 |

All the greatest men are maniacs. They are possessed by a mania which drives them forward toward their goal. The great scientists, the artists, the philosophers, the religious leaders—all maniacs. What else but a blind singleness of purpose could have given focus to their genius, would have kept them in the groove of their purpose? Mania, my dear Mister Bond, is as priceless as genius. Dissipation of energy, fragmentation of vision, loss of momentum, the lack of follow-through—these are the vices of the herd.

—Ian Fleming, *Dr. No* (1958)

## CUBAN MISSILE CRISIS

As the movie version of *Dr. No* was released on October 5, 1962, the Cold War had reached its peak. A mere eleven days after the first Bond movie hit our movie screens, the Cuban Missile Crisis struck. Also known as the October Crisis of 1962, the missile scare between October 16 and 28 was a thirteen-day toe-to-toe standoff between two superpowers: the United States and the Soviet Union. The crisis began with American ballistic missile deployment in Italy and Turkey. Consequent Soviet missile deployment in Cuba led to a confrontation often considered the closest the Cold War world came to escalating into a full-scale nuclear war.

The theme of this first Bond movie wasn't missiles, but rockets. The rather novel idea of stealing or toppling them. In Doctor Julius No, the world had its first Bond movie villain. For Doctor No had an island, as many devilish Bond movie scientists did later, along with his nuclear facility capable of producing an energy beam to topple unsuspecting American rockets on the launchpad. Conveniently for the contemporaneous cultural impact of *Dr. No,* the Cuban Missile Crisis also featured nail-biting intrigue about missile silos and secret bases.

*Dr. No* was made on a low budget of just over $1 million. Nonetheless, the movie turned out to be a financial success in the heightened Cold War climate. Over the years, the film has come to look rather dated, and indeed garnered mixed critical reactions upon release. The producers of the Bond movies started off with Ian Fleming's sixth novel for good reason: by the time of *Dr. No,* Fleming's stories had become increasingly elaborate and the villains flamboyantly megalomaniacal. Perfect for spy-fi.

## SEX, SNOBBERY, AND SADISM

Spy-fi's sub-genre cousin sci-fi had always suffered from a form of literary snobbery. For example, American science fiction writer Kurt Vonnegut had never been happy with the label "science fiction writer," as he was well aware that so-called "serious" literary critics used the genre as a "urinal," as he rather bluntly put it. So very few of Vonnegut's novels could be classified as straight sci-fi, though his fiction was always highly speculative. The same critical snobbery was shown to the spy-fi nature of *Dr. No* on its release in 1962. For example, on a weekly arts program broadcast on the BBC's Home

Service (now BBC Radio 4), which attracted an audience of several million, that week's program chair was a tweedy middle-aged critic by the name of Walter Allen. At one point in the program the playwright John Bowen, the youngest of the six program participants, told the other panelists they were "being rather patronizing" about *Dr. No*. After the slightest of pauses, Allen replied with a sigh. "Well, if you can't patronize Ian Fleming, who *can* you patronize?" (The British journalist Paul Johnson had started it all, with his famous 1958 *New Statesman* review of Fleming's novel, *Dr. No*, entitled "Sex, Snobbery and Sadism." Johnson wrote that the key ingredients of the novel were "the sadism of a schoolboy bully, the mechanical two-dimensional sex-longings of a frustrated adolescent, and the crude snob-cravings of a suburban adult." The review had soon got Johnson into trouble when he found himself "sitting at dinner next to Annie Fleming, who was Ian Fleming's wife, and she gave me a tremendous [ticking off] and rapped me over the knuckles with her spoon. I thought she was well suited to be married to the creator.")

## THE BOND FRANCHISE

And yet over time *Dr. No* has won a reputation among some as one of the Bond movie franchise's best installments. And what a franchise it is. Adjusted for inflation, Bond is the highest-grossing movie franchise of all time (and the first saga to reach $10 billion of grossing; for more data see IMDb):

1. *James Bond*
2. *Star Wars*
3. *Marvel Cinematic Universe*
4. *Harry Potter*
5. *The Lord of the Rings*
6. *Batman*
7. *Jurassic Park*

8. *Spider-Man*
9. *Pirates of the Caribbean*
10. *X-Men*

Not only that, but in 2003, Bond, as portrayed by Connery in *Dr. No*, was chosen as the third-greatest hero in cinema history by the American Film Institute, behind Indiana Jones in *Raiders of the Lost Ark* and Atticus Finch from *To Kill A Mockingbird*. (To get an alternate perspective on the potency of Bond, the third-greatest villain in cinema history to be chosen was Darth Vader in *The Empire Strikes Back*.)

And so *Dr. No* marked the first of a successful series of twenty-four Bond movies, and launched the genre of "secret agent" films, which thrived in the 1960s. The film also spawned a comic book and a soundtrack album as part of its PR and marketing. Of course, when *Dr. No* was released, the Bond series was not a franchise. That term wouldn't surface in relation to cinema for decades to come. Nor, in those early days, was Bond an institution, or a belated source of British self-esteem, as some people have tried to suggest. There's little evidence that *Dr. No* is some kind of shameful antiquated badge of national pride, of a previously "Great" Britain trying to be seen as punching above its weight on the global stage.

## THE PLOT AND BOND STYLE UNFOLD

Instead, *Dr. No* is a modest thriller. Bond is portrayed by Sean Connery, a handsome Scot who had, like Bond, spent some time serving in the Royal Navy. Connery's Bond is a rugged but dapper hero, with more charm than mind-boggling tech, who does his job without the need of spookish ordnance. In his first starring role, Connery is self-assured without being arrogant, a secret agent at ease in a dinner jacket but not born to the purple.

Some of the standard Bond tropes are present in this opening movie. As the film credits roll, they introduce the now-familiar Bond theme, with our hero firing his Walther PPK straight down the camera lens, the larger-than-life spy-fi villain with plans of world domination, the Bond girls to be bedded, and the wry comments delivered at the deaths of enemies and villains. And yet, at this early date, the tropes are newly born, less self-aware, and even arguably innocent, barely stamped by the Bond brand as yet, with no patented labels announcing their proprietary status.

Incidentally, we can also trace some of the Bond elements to their origin through the words of award-winning Bond set designer Ken Adam. In an interview with *The Guardian* in 2002, Adam talked about the inspiration behind his spectacular set for *Dr. No.*

> When they presented me with the first one hundred pages of *Dr. No*, which was pretty piss poor, I decided to do it because I knew Cubby Broccoli, and Harry Saltzman, and Terence Young, the director. We had a limited budget and everybody was in a hurry. It was unbelievably successful, which nobody had expected. There was a set that I designed as an afterthought, which I called the Tarantula Room. I don't want to bandy figures about, but I think it was about £450 that I had left to do this set. I had to design and build it very quickly. It had a circular skylight and was built in a false perspective. There was nothing in it except a chair, the door, and in the foreground a table with a tarantula cage. And so, you got this surrealistic, very simple set with an incredible effect. This set maybe was responsible for setting the style of future Bond films. I think that in terms of simple stylization it was one of the most effective I ever did.

The plot of *Dr. No* begins with the killing of the British MI6 station chief in Jamaica. He and his secretary are ambushed, and the assassins remove files on "Crab Key" and "Doctor No." In retaliation, the now-familiar M, the head of MI6, sends Bond out to investigate whether this new case is linked to another of Bond's cases; i.e. his collaboration with the American CIA on the disruption by radio jamming of rocket launches from Cape Canaveral.

On his arrival in Jamaica, Bond meets Felix Leiter, a CIA agent on the same mission as Bond who will become a regular feature in Bond films, played by a rather confusing number of different actors, making one wonder whether Bond's faculties as an agent fail to include that of face recognition. It seems the ever-resourceful CIA have traced the general whereabouts of the radio jamming signal to Jamaica, but have yet to discover its precise origin. Just as well Bond is here. They soon discover that, of the nearby islands, the reclusive Doctor No is the owner of Crab Key, an island fiercely protected against intruders by a private security force.

Using a Geiger counter, that well-known instrument used for detecting and measuring radiation, Bond discovers that mineral samples from the local islands are strong emitters of radiation, and convinces his reluctant crew to take him to Crab Key. There, in a now iconic scene, Bond meets the beautiful, bathing-costumed Honey Ryder, played by famous Swiss actor Ursula Andress. Ryder is an avid collector of shells, of the marine rather than ordnance kind. She leads Bond inland to a swamp area replete with radiation. It's here that they are attacked by one of the most curious pieces of Bond tech, the so-called "dragon" of Crab Key, a legend that has kept many a local away, but which turns out to look more like a shark, and to be nothing more than a creatively decorated and flamethrower-equipped tank.

## DOCTOR JULIUS NO

To set up the inevitable confrontation between Bond and his quarry in *Dr. No*, Bond and Ryder are kidnapped and end up in Doctor No's secret base. In Fleming's novel, Julius No confesses to Bond his cunning plan involving the interference of the rocket-testing center, boasting, "Their rockets would go mad. They would land on Havana, on Kingston. They would turn round and home on Miami . . . there would be panic, a public outcry. The experiments would have to cease . . . bases would have to close down."

Doctor No turns out to be a German-Chinese malevolent boffin, whose two prosthetic metal hands, due to radiation exposure, pile on the evil scientist tropes. (By the way, the word "boffin" is an informal name, mainly British, which is given to scientists who know much about science but little about ordinary life; those things that make us more human.) With delicious spy-fi drama, Doctor No reveals that he stole ten million dollars from a Chinese crime tong, a type of organization found among Chinese immigrants living in the US.

There are at least four strikes on the dubious nature of the character of the scientist Doctor No. **Strike One:** anything Chinese was pretty dodgy back in the day, as only in 1949 China had followed Russia and had one of those pesky communist revolutions. **Strike Two:** Doctor No is sinister enough to have masterminded the theft of a huge amount of money from a Chinese crime syndicate . . . he's smart enough to have outwitted the Chinese? **Strike Three:** Doctor No is now closely involved with a secret organization that goes by the laughable name of SPECTRE (**SP**ecial **E**xecutive for **C**ounter-intelligence, **T**errorism, **R**evenge, and **E**xtortion). These days, of course, criminal networks tend not to so blatantly confess by acronym the true intent of their "secret" organizations.

## BLACK-GLOVED BOFFINS

And yet the most revealing aspect of Doctor No's nature comes with **Strike Four:** his two black gloves. It's long been the case that writers and moviemakers will use certain motifs to signify something slightly strange or "other" about a character. And movie scientists get a special kind of treatment. Cary Grant is made to wear Coke-bottle glasses to play a mild-mannered paleontologist way back in the 1938 movie *Bringing Up Baby*. His specs are an exterior sign of interior issues. They imply the scientist doesn't see things like the rest of us. His myopic condition also cuts him off from some elements of the real world. In this way the filmic choice of simple prop can infer a symbol of separateness from the emotional depth the rest of us, hopefully, enjoy, rather than the obsessive focus of someone whose vision is limited to that of pure intellect alone.

The story of Doctor No's black-gloved hands crystallized with Fritz Lang's seminal film *Metropolis*, made in 1926. The movie's famous evil genius scientist was Doctor Rotwang. Fritz Lang's belief was that "an audience learns more about a character from detail and décor, in the way the light falls in a room, than from pages of dialogue." And what devilish detail did Rotwang possess? A metal hand. A relic of some dark accident that had left him maimed. An emblem of his struggles with dark science, but also of his twisted mental powers. This kind of recipe was taken on board in the portrayal of later cinematic scientists: unnamed scars, withered limbs, crazy eyes, curiously outsized foreheads, and, naturally, the wheelchair. The cliché becomes quite ridiculous at times. Take the scientist figure in *Mad Max Beyond Thunderdome*, for example. This character is a dwarf, sitting on the shoulders of a giant. Now, this is probably a reference to Isaac Newton's famous disingenuous quote, "If I have seen further, it is by standing on the shoulders of giants." In *Mad Max*, the brain has become detached from the body, as it were. And yet Newton had been referring to the shoulders of fellow scientists such as Descartes, Galileo, and the ancients, rather than some ripped jock on whose shoulders he could be carted about. As far as cinema goes: witches have tall pointy hats and broomsticks, boffins have disabilities.

This trail leads directly to the door of Doctor Julius No. Doctor Strangelove had just the one gloved hand, which spastically jerked into a Nazi salute at the mere mention of the word "slaughter," but Doctor No managed *two* black prosthetic hands. How badass is that? The hands, an outcome of some dubious nuclear accident, gift Doctor No some uncanny powers; we see him over dinner with Bond crushing a solid metal statue with his "bare" right hand. *Uncanny* is a curious word in science. Sigmund Freud had used the title, "The Uncanny," for his 1919 essay about the specific psychological fears associated with missing parts of the human body. Freud wrote, "Dismembered limbs, a severed head, a hand cut off at the wrist . . . feet which dance by themselves—all these have something peculiarly uncanny about them, especially when, as in the last instance, they prove capable of independent activity in addition." While it's true that Freud's examples are drawn from a more ancient fantasy associated with witchy folk tales, his analysis still applies to modern fantasies, such as science fiction and spy-fi. The black-gloved hands of Doctors Strangelove and No conjure up infantile fears and rather unhomely histrionics. After all, the German word "unhomely" is also used for "uncanny." For Freud, "the severed hand has a particularly uncanny effect." And so we see the same game at play with Doctor Rotwang, whose metal hand had replaced his severed one, with Doctor Strangelove, whose whole arm is now "capable of independent activity," and finally with Doctor No, whose hands are actually severed by the tongs in Fleming's novel, and replaced by metal manual prostheses in the movie, but by simple pincers in the book.

Julius No plans to disrupt the Project Mercury space launch from Cape Canaveral with his radio beam. After trying in vain to recruit Bond into SPECTRE, Doctor No holds him prisoner in a cell. But Bond escapes, naturally, crawling through an air vent, dressing up as a mere worker, and making his way to Doctor No's control center that contains a nuclear pool reactor. As we see the American rocket launch, Bond overloads the reactor and pushes Doctor No into the reactor pool, sending the rest of his body the same way as his emaciated hands. (In the book, Bond eventually killed No by suffocating him in a mound of guano, giving a new meaning to the idea of someone being "batshit crazy.") Finally, as becomes cliché in later movies, Bond and Ryder escape the island in a boat as the entire devilish Doctor's lair is annihilated in an explosion. Conveniently for Bond's romantic interests, the boat runs out of fuel, and they are rescued by Leiter. But, as Bond and Ryder kiss, Bond lets go of the ship's tow rope to carry on with unfinished business.

James Bond had arrived. The world's best-known secret agent came at just the right point in time and was precisely what people of the early 1960s wanted. Or as Terence Young, the director of *Dr. No*, once put it, the movie's release seemed to have hit "not only the right year, but the right week of the right month of the right year." After this auspicious start in the movies for Bond, it's perhaps little wonder that in *Everything or Nothing*, 2012's official documentary on the 007 franchise, John F. Kennedy is quoted as saying, "I wish I had had James Bond on my staff."

# FROM RUSSIA WITH LOVE (1963)
# BUMPING OFF BOND
# AND KILLING CASTRO

| *From Russia with Love* | Year: 1963 | Bond: Sean Connery | Director: Terence Young |
|---|---|---|---|
| Producer: Albert R. Broccoli & H. Saltzman | Screenplay: R. Maibaum & B. Mather | Distributor: United Artists | Release Date: October 10 |
| Running Time: 115 minutes | Budget: $2 million | Box Office: $79 million | Body Count: 27 |

In the center of Bond was a hurricane room, the kind of citadel found in old-fashioned houses in the tropics. These rooms are small, strongly built cells in the heart of the house, in the middle of the ground floor, and sometimes dug down into its foundations. To this cell the owner and his family retire if the storm threatens to destroy the house, and they stay there until the danger is past. Bond went to his hurricane room only when the situation was beyond his control and no other possible action could be taken. Now he retired to this citadel, closed his mind to the hell of noise and violent movement, and focused on a single stitch in the back of the seat in front of him, waiting with slackened nerves for whatever fate had decided for B. E. A. Flight No. 130.

—Ian Fleming, *From Russia, with Love* (1957)

## SPUTNIK LAUNCHES THE SPACE AGE
The James Bond franchise got off to a sterling start. Critics have suggested that Ian Fleming's book of *Dr. No*, published in 1958, was a prescient foretelling of the Cuban Missile Crisis that ran

concurrently with the theatrical release of the *Dr. No* movie. That would have been *some* world-shattering PR stunt to have pulled off.

Fleming's *Dr. No* novel was published in a late 1950s that saw the wake of the scandal of the first Russian space satellite, Sputnik, in October 1957. Sputnik ushered in the space age, launched the space race, and heightened the Cold War. It also sent shock waves across America. Future President Lyndon B. Johnson seemed to dream like a Bond villain when he declared in a speech only months later that "Control of space means control of the world . . . there is something more than the ultimate weapon. That is the ultimate position—the position of total control over earth that lies somewhere in outer space . . . " The *Manchester Guardian* in Britain focused sharply on underlying American fears: "The Russians can now build ballistic missiles capable of hitting any chosen target, anywhere in the world." American paranoia peaked when it was realized that Sputnik had flown over the US not once but four times during its brief voyage.

So, in *Dr. No*, Fleming had seemed to predict the next Cold War crisis. Aspects of the coming Cuban conflict were to be found in his story, set in the Caribbean, where a mad Chinese scientist has a dastardly plan to mess with US rocketry using atomic-powered radio beams. The parallels of fact and spy-fi are easily drawn: a quasi-commie outfit is threatening America from an island in the Caribbean, even if it is Crab Key, rather than Cuba. The very fact that Bond seemed so close to such real-world conflicts made his character seem cooler still—a secret agent with his finger on the pulse of global events—and it was a crucial factor in Bond becoming as popular as he did.

## BOND GETS WHITE HOUSE BLESSING

In this Cold War climate, little wonder that a Bond story including Russia in the title was chosen next. Jacqueline Kennedy is said to have gifted John F. Kennedy his first James Bond book and, allegedly, the president was hooked from then on. Witness the fact that, at one point, Kennedy included Fleming's *From Russia with Love* as one of his top ten favorite books. And the 1963 movie adaptation of the book is said to have been the last film Kennedy watched in the White House before that final and fatal trip to Dallas, Texas. Now, curiously, Fleming had previously been invited to a White House dinner and while there had given Kennedy some dubious advice on how to defeat Cuban leader Fidel Castro. The Bond author is said to have preferred the following tactic: simply get a scientific announcement from American doctors stating that men's beards can attract radioactivity, and thus cause sterility. The result? This cunning propaganda would persuade the gullible Castro (one assumes in this American scenario that the Cubans have no advising scientists of their own able to see through the bullshit) to shave off his beard and, like Samson in the Book of Judges from the Hebrew Bible, Castro would lose his power, once his followers saw him as a regular person (another implicit assumption in this strategy is that regular persons clearly don't have beards).

## THE RUSSIAN PLOT

The movie version of Fleming's *From Russia with Love* is as austere as *Dr. No*. Once more we find a film laced with less extravagance, packed with less action, and kitted out with fewer tech gadgets than the films that followed. For example, like the book before it, the movie features an attaché case complete with ammunition, a throwing knife, an AR7 folding sniper's rifle, fifty gold sovereigns, and a tear gas cartridge disguised as a tin of talcum powder, held in place by cunning use of magnets. That's some attaché case! The relative absence of the Bond formula in hindsight should not fool us. Look what English film critic Penelope Houston wrote at the time: "The success of *Dr. No* has no doubt given the James Bond team added confidence, if that was necessary, and *From Russia with Love* is made by people who clearly know that they now have a gilt-edged formula to play with."

And yet, in many ways, *From Russia with Love* is even more downbeat than *Dr. No*. The movie wasn't actually based in Russia, not in those Cold War days. Film production was taken elsewhere: the opening and closing sequences are set in Venice, with interiors shot in Pinewood, England, and the Balkan train journey that climaxes in a truck-versus-helicopter shootout was filmed in Scotland (the idea for the scene of the bright yellow chopper dive-bombing Bond's yellow flower truck is possibly borrowed from Hitchcock's 1959 thriller, *North By Northwest*, in which Cary Grant's character gets similar treatment to Bond). Spoiler: a spare chopper was hard to come by in those days, and the film's special effects crew were almost arrested trying to purloin one at a local airbase. So, the chopper chase was filmed instead using a miniature radio-controlled helicopter. Why not take a look at the movie again and see if you can spot the fake?

The movie's main location is a very gloomy and unspectacular-looking Istanbul. Fleming's original story of *From Russia with Love* is a simple plot to seduce and honey-trap Bond, a plot carried out by Soviet undercover agency SMERSH. (It was Joseph Stalin himself who had coined the СМЕРШ ["SMERSH"] brand—a portmanteau of the ominous phrase Смерть шпионам, "Death to Spies." The agency was set up to combat Nazi spies infiltrating the Russian military, but in the Cold War that followed, the agency's brief was quickly broadened with the task of finding and eliminating any subversive elements, presumably including Bond.)

## NOT TAKING SIDES

The *From Russia with Love* film was not so anti-Russian, however. It's worth noting that 1963 also saw the publication of British spy author John Le Carré's *The Spy Who Came in from the Cold*. Like Ian Fleming, Le Carré had been an insider, working for both the

Security Service (MI5) and the Secret Intelligence Service (MI6). Le Carré's book was a revolutionary and game-changing espionage novel. Its moral presentation was that of intelligence services of both Eastern and Western nations engaging in the same expedient amorality in the name of national security, or of bankrupt ideologies. Le Carré's Western spies were also depicted as morally burnt-out cases.

Likewise, the movie of *From Russia with Love* chose to position itself in a somewhat ambiguous position. The villains now belong not directly to SMERSH, but to that other shady Bond organization, SPECTRE, to whom the key Russian operatives have secretly switched allegiance. The motley crew have some memorable misfits among their number, notably Colonel Rosa Klebb. (Readers may recall Klebb as the Bond villain who rather iconically attempts to kill Bond using the poison-tipped daggers hidden in her shoes.) Klebb manages to recruit, with much early 1960s repressed shoulder-stroking and knee-fondling, a beautiful Russian patriot by the name of Tatiana (she will become the honey trap for Bond), who is working at Istanbul's Soviet embassy. Tatiana is naive enough to still believe that Colonel Klebb is working for Mother Russia. The rather unlikely but intended sting is this: Tatiana will get in touch with British intelligence, and claim to have fallen in love with James Bond from his file photo (high romance was so easy in those days). As a result of her Bond-smitten state, Tatiana would, for no obvious reason, offer the British a Russian coding device, the Lektor machine, which could be used to decipher top secret messages from the Russians. SPECTRE would then use Bond's conspicuous presence in Istanbul to ignite Anglo-Russian tensions. Key killings would be blamed on the British and, by besmirching his reputation with secretly shot sex footage of his bedroom action with Tatiana, eliminate hated enemy Bond himself.

## "DEATH TO CASTRO"

Now, just in case we think such spy-fi is too far-fetched, let's take a moment to compare the attempted plots on Bond's life with developments back in the real world. The American political class had its own version of the SMERSH motto, but it wasn't so much "Death to Spies" as "Death to Castro." In November 2006, the UK's Channel 4 broadcast a documentary film called *638 Ways to Kill Castro*. The film told the story of some of the more bizarre attempts of the American CIA to kill the Cuban leader over a period of half a century. A similar time period, in fact, in which villains have been trying to bump off Bond. Six hundred and thirty-eight may sound like a staggeringly high number of attempted assassinations, and yet clearly the CIA were as keen on killing Castro as spy-fi villains were on taking out 007. Or, as former head of the US interests section in Havana, Wayne Smith, put it, Cuba had the same effect on the US that a full moon has on a werewolf.

*638 Ways to Kill Castro* showed numerous mad methods of assassination, straight out of the Bond franchise: from exploding cigars to femmes fatales. On one occasion, a radio station was rigged with a gaseous form of LSD (this sounds like a similar plan to that unleashed in the 2012 movie *The Amazing Spider-Man*, in which the Lizard plans to make all humans lizard-like by releasing a chemical cloud from Oscorp's tower). On another occasion, a poison syringe was the plan to be used, posing as an innocent-looking fountain pen.

All would-be assassins, the film claimed, were supported and funded by the US government. At one point the CIA even sought the help of the Mafia, hoping they would succeed where so many others had failed. The retired chief of Cuba's counterintelligence, Fabian Escalante, was the man tasked with the responsibility of protecting Castro for many of the almost fifty years Castro was in power. Seems like Escalante did a good job, or else American assassins aren't what they might be. Of the hundreds of plots dreamt up to end Castro's life, some were carried out directly by the CIA, especially during the Sean Connery Bond years of the first half of the 1960s. From the Roger Moore Bond years of the seventies onward, Castro assassination attempts were more often made by Cuban exiles, who had been trained by the CIA shortly after Castro took power in the Cuban Revolution of 1959.

## BUMPING OFF BOND

The James Bond franchise has a long history of trying to bump off Bond by scientific or technical means. In *From Russia with Love*, the attempt on Bond's life is carried out by a time device. Here, the blond SPECTRE killer named Grant, played by Robert Shaw, almost steals the picture. Earlier in the movie Klebb had established Grant's hard-man credentials by smashing him in the stomach with some knuckledusters (spoiler: he hardly flinches). On a train journey later in the film, Grant tries to strangle Bond with a wire hidden in his wristwatch. Readers may recall that the zoological attempt on Bond's life in *Dr. No* had been by a judiciously placed tarantula as Bond slept (incidentally, Bond kills the spider in time to the jabbing violin-stringed soundtrack, a passing cinematic nod to the shower murder in *Psycho*, which had been released two years earlier).

In the short time to *Goldfinger*, assassination attempts on Bond had already branched into more technical aspects of physics: 007 is plonked onto a plinth and a high-powered laser slowly approaches Bond's expectant crotch (this laser was so powerful that, according to Goldfinger, "It can project a spot on the moon—or, at closer range, cut through solid metal." By the way, this iconic scene also contains this wonderful exchange:

**Bond:** Do you expect me to talk?
**Goldfinger:** No, Mr. Bond, I expect you to die!

While *Thunderball* sees attempted murder by the technology of a motorized traction table, and *On Her Majesty's Secret Service* by hand grenade on a bobsled chase, *You Only Live Twice* treats us to two attempts, one by airplane crash (rather mundane) and another more chemically elaborate attempt by poison dripped along a hung cotton thread as Bond sleeps below. In *Diamonds Are Forever*, they apply a thermodynamic approach of cremating Bond alive in a coffin!

Indeed, the Bond murder attempts are almost as numerous and variable as the catalog of ways they tried to kill off Castro. When Roger Moore became Bond, the pace simply didn't let up. Assassination attempts by biological means are featured in the first three Bond movies that Moore did: by snake, then by crocodile and alligator in the New Orleans swamps in *Live and Let Die;* by a three-nippled assassin in *The Man with the Golden Gun;* and by being dumped out of an elevator into a shark tank in *The Spy Who Loved Me.*

They tried a snake once more on Bond in *Moonraker*, to little effect, then decided to delve back into physics by, in turn, using a Drax Industries gravity-force simulator; throwing him off a cable car roof in Rio (this by the infamous Jaws); and finally by exposing him to the rocket fuel thrust of Moonraker itself. In *For Your Eyes Only*, Bond is plummeted from an airborne chopper above London's Big Ben, then dragged through the ocean like a torpedo! In *Octopussy*, they tried dropping a revolving circular saw onto his head as he lay on the bed with Octopussy, and in *A View to a Kill*, dragging him over the Golden Gate Bridge as he dangled from a zeppelin, and then attacking him with a red axe. It's little wonder Roger Moore retired from the role.

## DARK UNDERWORLD ASSASSINS

Timothy Dalton got off relatively easy. In *The Living Daylights*, they tried the (very British) assassination of simply shooting Bond through the canvas roof of a Land Rover in a car chase on Gibraltar, then upped their game for *Licence to Kill* by firing a rocket into an oil tanker Bond just happened to be driving. The various attempts at assassination seem to have done little to persuade Bond to change the way he lives his life. The same could not be said for Castro. In his early years in office after the revolution, Castro would often walk alone in the streets of Cuba. But the CIA changed all that. They approached dark underworld assassins to carry out the most casual of killings. One of Castro's old classmates planned to shoot him dead in the street in broad daylight, much in the manner of a Mafia hitman. One would-be sniper at the University of Havana was caught by Castro's security team. And yet the shooters were no more successful than the poisoners and bombers. After a while, Castro doubles were used, and around twenty different addresses in Cuba were employed to make it harder for any potential hitmen to get at him.

The Pierce Brosnan Bond period saw a huge technical peak in assassination attempts. In *GoldenEye*, Bond has heat-seeking missiles launched at the very helicopter he's sitting in; get out of that one, Fidel! Later, Bond villain Sean Bean attempts to drop Brosnan off the Arecibo Radio Telescope dish in Puerto Rico. In *Tomorrow Never Dies*, Bond is almost strangled from behind in a MiG fighter, and later almost has his chakras skewered, while in *The World Is Not Enough*, Bond is almost sliced in two by rotating blades hung from choppers, then is subjected to an ordeal in a medieval torture chair, which works by a kind of capstan's wheel at the back that, when rotated, tightens the collar around the neck until the spine is plunged into the neck. It might seem a rather obvious question, but wouldn't it be better and quicker simply to shoot Bond in the head? You may remember a satire of this situation in *Austin Powers: International Man of Mystery*:

> **Dr. Evil:** Scott, I want you to meet daddy's nemesis, Austin Powers.
> **Scott Evil:** What? Are you feeding him? Why don't you just kill him?

**Dr. Evil:** I have an even better idea. I'm going to place him in an easily escapable situation involving an overly elaborate and exotic death.

No one could get close enough to Castro to pull off a torture chair challenge or, for that matter, a simple shot to the head. On one occasion, when a former lover did get close enough to Fidel to assassinate him, she gave up. She was recruited to kill Castro by using poison pills provided by the CIA. She hid them in her cold cream jar. But the pills melted in the Cuban heat and she decided that, all things considered, popping cold cream and pills in Castro's mouth while he slept was a bad idea (the technique has faint echoes of the drip-poison that villains tried on Bond in *You Only Live Twice*). According to this woman, Castro had guessed that she was primed to kill him, and helpfully offered her his own pistol. "I can't do it, Fidel," she told him.

## METHODS OF MURDER

Daniel Craig continues in the highly technical vein of Pierce Brosnan. In *Casino Royale*, Bond is poisoned during a poker match, then goes into cardiac arrest, before using a defibrillator secreted in his Aston Martin (spoiler: Bond is essentially dead for a few seconds, until Vesper presses the relevant red button to defibrillate our hero). In the same movie, a body is left in the middle of a rural road at night so that Bond's Aston Martin flips into many dangerous somersaults. And, if that isn't enough, later in the movie there's a knotted rope attack to the testicles (it's unclear whether this is life threatening, though it's at least guaranteed to make all male readers flinch). In *Quantum of Solace*, Bond has his airplane shot down in South America where he lands, without parachute, in a reservoir-filled sinkhole in the Bolivian desert. Incidentally, there were two water-based kill attempts on Castro: (a) knowing his fascination for scuba-diving, the CIA invested in a large volume of Caribbean mollusks. The plan was to find a shell big enough to house a lethal quantity of explosives, which would then be painted in colors bright and alluring enough to attract the nerdy Castro's attention while underwater; and (b) one aborted plot also related to Castro's submarine activities involved a special diving suit infected with a fungus that would infect Castro with a chronic and debilitating skin disease. Aren't the people who dream up such methods of murder the most charming of humans?

In *Skyfall*, an attempt is made to kill 007 by, rather unbelievably, creating an underground explosion in the London Tube system so that a tube train appears through the huge resulting hole in a subterranean wall and descends upon Bond! And in *Spectre*, after trying to throw him off a speeding train, the movie later features a rejuvenated Blofeld who has diabolically created a technologically superior torture chair, resembling a cross between a dentist's chair and a robot, with which he attempts to drill Bond's skull. Perhaps all of these assassination attempts explain why, in the original book of *From Russia, with Love*, and as we quoted at the start of this chapter, Fleming wrote that "In the center of Bond was a hurricane room, the kind of citadel found in old-fashioned houses in the tropics . . . Bond went to his hurricane room only when the situation was beyond his control and no other possible action could be taken."

Maybe Castro too had a hurricane room. It's certainly true that jokes about Castro's apparent indestructibility became commonplace in Cuba. One such story told the tale of Castro being given a gift of a Galapagos turtle. Castro declined the gift when it was explained that the turtle would live only one hundred years. Castro is alleged to have said that the problem with pets was you got attached to them, and then they died on you.

# GOLDFINGER (1964)

# MAN HAS ACHIEVED MIRACLES IN ALL FIELDS BUT CRIME!

| Goldfinger | Year: 1964 | Bond: Sean Connery | Director: Guy Hamilton |
|---|---|---|---|
| **Producer:** Albert R. Broccoli & H. Saltzman | **Screenplay:** R. Maibaum, P. Dehn & B. Mather | **Distributor:** United Artists | **Release Date:** September 17 |
| **Running Time:** 110 minutes | **Budget:** $3 million | **Box Office:** $125 million | **Body Count:** 77 |

**Goldfinger** [pointing with the cue]: This is my bank. The gold depository at Fort Knox, gentlemen. In its vaults are fifteen billion dollars—the entire gold supply of the United States . . . Man has climbed Mount Everest. Gone to the bottom of the ocean. He has fired rockets to the moon. Split the atom. Achieved miracles . . . in every field of human endeavor—except crime!

 —R. Maibaum, B. Mather & P. Dehn screenplay, *Goldfinger* (1964)

**Pussy Galore:** My name is Pussy Galore.
**Bond:** I must be dreaming.

 —R. Maibaum, B. Mather & P. Dehn screenplay, *Goldfinger* (1964)

## THAT MIDAS TOUCH

*Goldfinger* was huge. The first real Bond blockbuster, *Goldfinger* had a budget equal to that of the first two films combined. A financial success, the movie's $3 million budget was recouped in only two weeks. Indeed, *Goldfinger* broke box office records in many countries around the world, as *The Guinness Book of World Records* later listed the movie as the fastest grossing film of all time. Demand for the film was so intense that the DeMille Theatre in New York City had to stay open twenty-four hours a day, as the movie's release led to a number of marketed and licensed tie-in items. As a boy of six years old, I owned a toy Aston Martin DB5 car from Corgi Toys, which was the biggest selling toy of 1964. (The toy DB5 had a tiny James Bond figure driving at the wheel, retractable machine guns, a roof that opened out with a functioning ejector seat, a rear bullet screen, and all contained in a fully assembled die-cast metal model packaged in a display box.) The movie's promotional tie-ins also included an image of gold-painted Shirley Eaton as Jill Masterson on the cover of *Life* magazine. *Goldfinger* was the first Bond film to win an Academy Award and played to an overwhelmingly favorable critical reception. In 1999, the film was ranked number seventy on the BFI Top 100 British films, in a list compiled by the British Film Institute. And all this resounding and glittering Midas-like success for a movie about gold.

With *Goldfinger*, Bond came of age. Now we begin to see some of the elements familiar to many of the later James Bond films. Bond's extensive use of gadgetry and technology starts to surface, along with multiple exotic locations, that playful, tongue-in-cheek humor, and an extensive pre-credits sequence that had little bearing on the main plot. (By the way, the very first *Goldfinger* gadget is a fake seagull on Bond's head. You read that right—this Bond blockbuster and original of 007 spy-fi films begins with Bond doggy-paddling underwater with his night-time presence disguised by a fake bird plonked on top of his head. Blink and you'd miss this seagull delight, as it soon segues into the now legendary wetsuit-gives-way-to-tuxedo scene.)

## ALL THAT GLITTERS

And that plot is all about the science and economy of gold. We find Bond in Miami Beach on vacation. He's told by his superior, M, via the CIA agent Felix Leiter, to investigate the bullion dealer Auric Goldfinger at the hotel there (the name Auric is derived from the Latin name for gold). The first altercation Bond has with Goldfinger doesn't end well. Bond is knocked out by Goldfinger's Korean manservant, Oddjob, and Bond wakes to find another Goldfinger employee, Jill Masterson, dead on the bed, having died from "skin suffocation" after being completely covered in gold paint. Bond says in the movie, "She died of skin suffocation. It's been known to happen to cabaret dancers. It's all right provided you leave a small bare patch at the base of the spine to allow the skin to breathe." (Incidentally, an urban legend spread that the actress who played Masterson, Shirley Eaton, actually died on set from asphyxiation from the gold paint. Eaton was at first reluctant to take the role when sent by her agent to meet producer Harry Saltzman, not because of the threat of suffocation, but because of the prospect of being naked on camera. She agreed to take the part only if the nudity was done tastefully. The production team took an hour and a half to apply the gold paint to Eaton's body, and I'm pleased to say Eaton is alive and kicking and in her early eighties as I write this.) The "skin suffocation" is a physical impossibility, as most humans breathe through their noses and mouths, and not through the skin. However, had Goldfinger arranged for the gold paint to have had some toxic elements in it such as lead or mercury, it's possible that Masterson could have been poisoned . . . slowly. The toxins would take time to penetrate her bloodstream. *Mythbusters*, the science entertainment television program, also looked into this matter back in 2009 and concluded that while death by suffocation wouldn't result, a painted person would nonetheless undergo dramatic shifts in their blood pressure and body temperature. Still, dramatic license, I guess.

Back in London, M and no less than the Chancellor of the Exchequer himself (this movie is very grand in its vision) think they know the science of what Goldfinger is up to. As the price of gold varies across the globe, a bullion dealer like Goldfinger can profit by smuggling gold and selling bullion internationally. Goldfinger is nothing more than a profiteering pirate!

## "EJECTOR SEAT? YOU'RE JOKING!"

Bond's mission, should he choose to accept it, is to find out all about Goldfinger's gold bullion smuggling operation, armed with his modified Aston Martin DB5. You may recall this is the Aston Martin with the celebrated ejector seat and the now famous exchange between Q and Bond that goes something like:

> **Q:** Now this one I'm particularly keen about. You see the gear lever here? Now, if you take the top off, you will find a little red button. Whatever you do, don't touch it.
> **Bond:** Yeah, why not?
> **Q:** Because you'll release [points out roof] this section of the roof, and engage and then fire the passenger ejector seat. Whish!
> **Bond:** Ejector seat? You're joking!

The legendary gadget-laden DB5 was introduced twenty-two minutes into the movie, leaving moviegoers with half an hour of bated breath until all the gadgets were put to use. Along with the ejector seat, the DB5 comprised a bulletproof body; on-board radar and tracer; smoke screen, oil slick, tire shredder, and machine gun functions; a bulletproof screen; and a set of revolving license plates. While many of these features would have proved impractical or impossible in real life, it didn't stop Corgi's toy DB5 from becoming a bestseller that year. The revolving plates were an understated advantage in the movie as Bond was about to travel from London across the channel to Calais, France, then, following Goldfinger, through Rouen and Orléans and finally to Geneva, Switzerland. Having a local number plate at each point would attract less attention for someone wishing to travel relatively incognito, as long as Bond didn't somehow set off the ejector seat in a built-up area.

Bond soon finds that Goldfinger smuggles his bullion by simply melting it down and hiding it in the bodywork of his Rolls-Royce Phantom III (how British is *that*?), which goes with Goldfinger everywhere on his travels (though the weight of the gold must surely play havoc with the Phantom's suspension and tires, and of course excessive vehicle weight would lead to road damage, other vehicle damage, accidents, injuries, and even death. But this is unlikely to worry a criminal brain like Goldfinger's, whose concentrated mind is focused solely on his cargo). Incidentally, any curious cop with an eye for chemistry may be alerted to the Phantom's number plate of "AU1," as Au is the periodic table chemical symbol for gold. A long shot, admittedly, but who knows?

(By the way, here's an experiment you can do at home. Why not rewatch the *Goldfinger* movie and make up your mind as to whether you think Donald Trump has based his style on Auric Goldfinger? We're not saying that Trump has plans to blow up the entire federal gold reserve in Fort Knox, but he does share Goldfinger's obsession with gold. Not only that, but it's hard not to notice that from certain angles there's an uncanny resemblance between the look of the Bond villain and the current US president. Of course, if you are convinced, depending on your political views, you may enjoy the movie a little less, but the converse effect also applies. Once you've been exposed to the Trump-Goldfinger illusion, every time you experience Trump in the future, you'll get a tad of Goldfinger, albeit on a very low, perceptual, and possibly unconscious level. This may also be a useful political ploy for Trump's opponents. All they need do is plant mistrust in the minds of Trump supporters.

This may, after all, work better than actually arguing clearly and rationally about his policy decisions.)

## OPERATION GRAND SLAM

More importantly, Bond discovers Goldfinger's "Operation Grand Slam." He hears Goldfinger talking to a Chinese nuclear physicist named Mr. Ling about the operation, then witnesses Goldfinger's meeting with American mafiosi, who have brought the materials he needs for the operation. Goldfinger reveals his plan to rob the US Bullion Depository at Fort Knox by releasing Delta 9 nerve gas into the atmosphere. (Goldfinger later has the mafiosi killed using the gas.)

Ken Adam designed the interior of Fort Knox for the movie and told *The Guardian* the following in 2002:

> The inside of Fort Knox in *Goldfinger* is completely unreal. There's no place like that, any more than the war room I designed for Kubrick's *Dr. Strangelove* exists. There was something somber in these sets, and the thing that must have influenced me was the German expressionism I grew up with—films like *Dr. Caligari* and so on. People point to my past—persecution by the Nazis and suchlike. I don't know. I know that I was influenced by expressionism, there's no question about it. But when you look at all these command centers and villains' bases, they're not based on anything— they're just from my imagination.

Bond tells Goldfinger that his plan to rob Fort Knox won't work ("$15,000,000,000 in gold bullion weighs 10,500 tons. Sixty men would take twelve days to load it onto two hundred trucks. Now, at the most, you're going to have two hours before the army, navy, air force, marines move in and make you put it back"), to which Goldfinger replies, "Who mentioned anything about removing it?" Rather, Goldfinger's plan is to detonate a dirty bomb inside the vault of Fort Knox, rendering the gold allegedly useless for fifty-eight years (a very precise figure indeed; one wonders why they didn't also add "and five months" or whatever). The result of the dirty gold would be to increase the value of Goldfinger's gold, therefore gifting the Chinese an advantage in the ensuing potential economic crisis.

But why has gold captured the human imagination for thousands of years? The rule of the pharaohs in Ancient Egypt conjures up gold in the minds of many people. Gold was actually quite scarce in Ancient Egypt, so the rulers kept the gold to themselves, Goldfinger style. They used it in their exquisite jewelry, as gifts to other monarchs, but mostly they used gold in their tombs and their funeral masks. Gold doesn't tarnish, and is pretty indestructible, so they associated it with eternal life. They felt it helped their journey from this world into the world of the gods. And the gods themselves had skin of gold, and bones of silver, they believed. As Goldfinger says, "This is gold, Mr. Bond. All my life I've been in love with its color . . . its brilliance, its divine heaviness."

## GOOD AND BAD GOLD SCIENCE

Goldfinger's "Operation Grand Slam" has good points and bad points. A good point is that gold *is* a scarce commodity. Up until the year 1850, scarcely ten thousand tons of gold had been mined in all of human history (that's roughly the same amount Bond thinks Goldfinger is about to steal from Fort Knox). One polar bear weighs around a ton, so it's the same weight of gold as ten thousand polar bears. That might sound like a lot (and ten thousand gold polar bears running about sounds pretty cosmic), but this is *all* of human history we're talking about. Put another way, a blue whale weighs about one hundred tons, so it's the same amount of mined gold as one hundred blue whales, in all of history. Yet another way of thinking about the scarcity of gold is realizing that only one hundred thousand tons have been mined since 1500. That's enough to fill only a sixty-foot-square crate that would sit neatly under the Eiffel Tower. (Sure, there's huge amounts of gold in the sea, where there are ten million tons of the stuff floating about: worth around $450 trillion, but the price of extracting anything at a concentration as small as ten parts per trillion are so high it'd even put Goldfinger off.)

A bad point of Goldfinger's "Operation Grand Slam" is his dirty bomb theory. Mr. Ling, the Chinese nuclear physicist, should have told Goldfinger that the damage done by such a bomb is twofold: the initial blast, followed by the resulting radiation. Small-time pulp-fiction villains might get their hands on such small nuclear devices, but Goldfinger didn't plan to blow anything up. Rather, he intended to use the so-called "dirty" effect of irradiating the gold. Mr. Ling might have said that, in theory, gold could have become radioactive from a cobalt-iodine atomic device (which Goldfinger claimed to have sourced from the Chinese) after gaining an extra neutron from the plethora of subatomic particles boomed out by the blast. But such a bomb wouldn't keep the US gold reserve untouchable for fifty-eight years. The radioactive form of gold is very unstable. It would turn to liquid mercury within days. Instead of making the Fort Knox reserve untouchable, Goldfinger would effectively annihilate it. Having said that, the net effect of his gold-into-mercury magic trick might still make his stock value go up by ten times, as Goldfinger says in the movie.

In the end, it's all rather academic as, naturally, Bond manages to cunningly electrocute Oddjob. Then, after troops finally kill all of Goldfinger's men and open the vault, an atomic specialist rushes in with seconds to spare and simply turns off a timer on the device. The time left before everyone is blown to kingdom come? Why 0:07 seconds, of course.

# THUNDERBALL (1965)
# MAKING A MINT OUT OF CRIME

| *Thunderball* | Year: 1965 | Bond: Sean Connery | Director: Terence Young |
|---|---|---|---|
| **Producer:** Kevin McClory | **Screenplay:** R. Maibaum & J. Hopkins | **Distributor:** United Artists | **Release Date:** December 9 |
| **Running Time:** 130 minutes | **Budget:** $9 million | **Box Office:** $141.2 million | **Body Count:** 110 |

**Blofeld:** [on tape recording] My dear Prime Minister, two atomic bombs, numbers 4-5-6 and 4-5-7, which were aboard NATO Flight 7-5-9, are now in the possession of SPECTRE. Unless within the next seven days your government pays to us one hundred million pounds sterling, in a manner to be designated by us, we shall destroy a major city in England or the United States of America. Please signal your acceptance of our terms by arranging for Big Ben to strike seven times at 6 p.m. tomorrow.

—R. Maibaum & J. Hopkins screenplay, *Thunderball* (1965)

I remain terrified of the capacity of the media, the capacity of spin doctors, here and abroad, particularly the United States media, to perpetuate false lies, perpetuate lies. Mussolini, I think, defined fascism as the moment when you couldn't put a cigarette paper between political and corporate power. He assumed, when he offered that definition, that media power was already his. But I worry terribly that the absence of serious critical argument is going to produce a new kind of fanaticism, the new simplicities that are as dangerous as the ones which caused us to march against Iraq and as misunderstood.

—John Le Carré, interview with *Democracy Now* (2010)

## SPY-FI BY DESIGN

*Thunderball* was meant to have been the first film of the Bond franchise but, for legal reasons over copyright issues, it wasn't made until 1965. While many critics and moviegoers showered the film with praise and dubbed it a welcome addition to the series, others pointed critically at the repetitively monotonous underwater action and the movie's prolonged length. Nonetheless, the film was a roaring success, making a total of $141.2 million worldwide, more than the earnings of the three previous Bond films. *Thunderball* is the most financially successful film of the franchise in North America, if adjusted for movie ticket price inflation.

The following year, the movie won the Academy Award for Best Visual Effects, with Bond production designer Ken Adam also being nominated for a BAFTA award. Born a German Jew in 1921, Klaus Hugo Adam had been one of only three German-born pilots to serve in the British Royal Air Force during WWII. (As Adam once confessed, "I was always enormously attracted to speed, adventure, and excitement, so certainly it somehow found a reflection in some of my Bond designs, without any doubt.") At the age of ninety-one, Adam was interviewed by freelance journalist Todd Longwell and spoke about the spy-fi visual template for the 007 franchise and his work on seven Bond films between *Dr. No* and *Moonraker*. When Longwell asked Adam why he thought his work had turned out so well, Adam replied:

> There was a lot of luck. When I designed *Dr. No*, everything sort of worked out right. The main unit was working in Jamaica and I was left to my own devices. The studio was here in England. I suddenly came up with the idea that it would be fun to, apart from the tongue-in-the-cheek element, to express our technological age, which I hadn't seen at that time, except going back to films like Fritz Lang's *Metropolis* in 1927. For many, many years, we used the same way of

building the sets and they were never really reflecting this neurotic technological age that we were living in.

*Thunderball* opens with a scene that pretty much captures what Ken Adam said to Todd Longwell. We see Bond escaping with the use of a jetpack, which once more firmly places Bond in spy-fi. The jetpack, or rocket-pack, is the concept of a flying device, worn on a person's back, which uses jets of liquid or gas to propel the jetpack pilot through the air. The idea had been pretty common in sci-fi for most of the twentieth century, but became particularly widespread in the 1960s. Real jetpacks have been developed using a variety of technologies. Alas, in practice, real jetpacks are far more limited than the fictional versions, such as we see in *Thunderball*. In reality, the challenges of the earth's atmosphere, gravity, the low energy density of potential fuels, and the human body not being suited to fly make it all too tricky and lead some to complain, "Dude, where's my jetpack?" Astro Teller, head of the Google X research laboratory, went on record in 2014 to say they'd investigated jetpacks and found them way too fuel-inefficient to be practical, with consumption as high as 940 liters per hundred kilometers. They were also as loud as motorcycles. So Google stopped developing them. These days jetpacks are usually used for stunts, such as the famous Bond scene in *Thunderball*, or in extra-vehicular activities for astronauts.

## SPECTRE

Soon after the jetpack scene, the action cuts to a SPECTRE meeting in Paris. The scene is again evidence of Ken Adam's genius. SPECTRE had been mentioned in *Dr. No* and *From Russia with Love*, but only now does Adam reveal the futuristic headquarters of the terrorist organization. In a now iconic scene, Bond's arch enemy, Blofeld, is half-hidden and sitting in a high-backed black leather chair stroking a white cat on his lap (this is, of course, one of the key iconic scenes amusingly satirized by Mike Myers in the *Austin Powers* series). In 2002, Adam described his set design to *The Guardian*:

> By this time I was pissed off with all these vast command centers and conference rooms, and so for the SPECTRE headquarters I decided to get away from the normal confer-ence-room concept. Instead of a table I had these chairs with individual control consoles facing each other in two lines. And then I had the idea of one of the chairs disappearing under the floor with a man who had betrayed SPECTRE and coming up empty. What I tried to introduce into these villainous sets was a bit of humor—they're not really real.

In mentioning those "vast command centers and conference rooms," Ken Adam is rather modestly referring to his own work as the creative artist for another iconic cinematic set design in *Dr. Strangelove or: How I Learned to Stop Worrying and Love the Bomb*. Released the year before *Thunderball*, *Dr. Strangelove* featured a nuclear-proof conference room underneath the Pentagon, the iconic War Room. As with *Thunderball*, the design leaned heavily on sets from German expressionist cinema, such as *The Cabinet of Dr. Caligari* and Fritz Lang's *Metropolis*.

In the SPECTRE meeting, the eye-patched Bond villain Emilio Largo reveals the group's latest project: the hijacking of two atomic bombs and the subsequent holding of NATO ransom. Soon we see the hijack, landing the atomic bombs in shallow Bahamian waters close to Largo's ship, the *Disco Volante*, whose SCUBA divers retrieve them. Ken Adam described the *Disco Volante* boat as a complete experiment. The production had needed a yacht for the villain that could do sixty knots and yet still look like a yacht. Such yachts didn't exist at the time. As luck would have it, a hydrofoil came up for sale, which the production team bought for $10,000. Adam redesigned the craft by extending the hull and making a catamaran.

## CARNIVAL BOND

The scene switches to another classic Ken Adam set. We're at an MI6 conference in London, often dubbed "M's Conference Room." The room is impressive in scope and purpose, large and elegant enough to accommodate heads of state: secure and high-tech, and yet also classically ornate, with tapestries and paintings that disguise a variety of maps, data screens, and status boards. It's the ultimate situation room, and now it's being used to house all nine members of the double-0 section in a major spy-fi crisis of world-wide implications. The agents are told that SPECTRE demands £100 million from NATO in exchange for returning the bombs (with the great dramatic flourish in another scene that the authorities should arrange for Big Ben to strike seven times at 6:00 p.m. to signal to SPECTRE that the blackmail is on). Otherwise, SPECTRE threatens to destroy a major city in the United States or the United Kingdom. Incidentally, it's worth noting that Emilio Largo in Paris and Bond in London are both late for their respective briefings at their headquarters, serving to underline that the two main players are equally matched in status and character.

Bond goes to Nassau in the Bahamas and is soon kidnapped, but escapes through a Junkanoo celebration (a street parade with music, dance, and costumes of Akan origin, which is ritually celebrated in many islands across the Bahamas). And so begins another Bond tradition, what we might dub "Carnival Bond." With immaculate good luck and fortuitous anthropological timing, there often seems to be a carnival of some sort handy when Bond needs desperately to make a getaway from his enemies. Whether it's the blues funeral and voodoo ritual in *Live and Let Die*, the spectacular Mexico City Day of the Dead parade in *Spectre*, the Swiss Christmas fayre in *On Her Majesty's Secret Service*, Rio's Mardi

Gras celebration in *Moonraker*, or the busy Indian market scene in *Octopussy*, whenever Bond needs to make a dash for it, local culture and commerce are obligingly and cacophonously at hand.

## THE SILENT WORLD

Bond ultimately discovers the bombs and Largo's plan to destroy Miami Beach, but not before a series of very orchestrated and choreographed underwater battles have ensued. The primary difficulty in underwater camera usage had been in sealing the camera from water at high pressure while trying to operate it. The first full-length, full-color underwater documentaries were developed in the mid-1950s. Jacques Cousteau's 1956 movie *The Silent World* is noted as one of the first films to use underwater cinematography to show the ocean depths in color. Its title derives from Cousteau's 1953 book *The Silent World: A Story of Undersea Discovery and Adventure*. Cousteau went on to make some of the world's greatest underwater videos, which inspired thousands of deep-sea explorers and keen wildlife moviemakers. Cousteau's work meant that, for the first time, many people saw new species of flora and fauna, famous historical shipwrecks, and unexplored underwater caves. Cousteau's films gave underwater filmmaking greater prestige and attention, and it's exactly this cultural cachet that *Thunderball* draws from. A year after *Thunderball* was released, American television companies created the documentary television series *The Undersea World of Jacques Cousteau*, which ran for ten years from 1966 to 1976.

## CRIME PAYS

Bond succeeds in his mission to find the two NATO atom bombs stolen by SPECTRE, which had held the world ransom for £100 million in sterling. According to British spy author John Le Carré, in reality the authorities are far more complicit when dealing with large criminal organizations. And Le Carré should know; he's worked for both MI5 and MI6, and has been keeping tabs on global espionage for over sixty years. In an interview with *Democracy Now* in 2010, Le Carré said, " . . . we're up against the fact that huge corporations are effective here, control the super markets, whatever they do, and they pay virtually no tax." Le Carré then focuses on the role of the spy agencies: "And what we are gradually learning from these little exposés that come to light is that there is almost no way of denying people, in the end, the profits of their crime, which is a tragedy." So, who needs the fictional fantasy crimes of SPECTRE? Real-life crime pays. And as John Le Carré concludes, that's not good news for the great majority of us.

# YOU ONLY LIVE TWICE (1967)
# THE RACE TO
# CONQUER SPACE

| *You Only Live Twice* | Year: 1967 | Bond: Sean Connery | Director: Lewis Gilbert |
|---|---|---|---|
| Producer: Albert R. Broccoli & H. Saltzman | Screenplay: Roald Dahl | Distributor: United Artists | Release Date: June 12 |
| Running Time: 117 minutes | Budget: $10.3 million | Box Office: $111.6 million | Body Count: 196 |

You only live twice: once when you are born, and once when you look death in the face.

—Ian Fleming, *You Only Live Twice* (1964)

We all think we know the story of space—that it was conquered by the Americans. But that is not the real story. The actual conquerors of space were a group of men, and women, from the other side of the Iron Curtain. After decades of secrecy, they are now free to tell the extraordinary stories of how they risked everything to take the first steps into space. This is the remarkable, and at times terrifying, story of the cosmonauts.

—Michael Lachmann, *BBC's Cosmonauts: How Russia Won the Space Race* (2014)

## BIG IN JAPAN
The Bond franchise skipped the year 1966 and jumped straight into the summer of love that was 1967 with *You Only Live Twice*. The plot, in short, goes something like this: Bond goes to Japan

and discovers that Blofeld has built a super sci-fi spaceship that has been kidnapping (yes, kidnapping!) the meager spaceships of the US and the USSR. And how does Blofeld manage to pull off this fantastic yet covert operation? By launching his super-spaceship from a Japanese volcano base, naturally. So, putting to one side the space race between Russia and America, Bond has bigger fish to fry.

Looking at this brief plot outline, it's little wonder that many regard this movie as the ultimate spy-fi Bond film. It manages to pull off some really ambitious science fiction ideas without plummeting into parody (Mike Myers took *that* final step with *Austin Powers*, of course), which maybe makes this the movie from which most of the Bond archetypes stem.

The film of *You Only Live Twice* ignores almost all of the original novel. Fleming's book had been a bitter tale of a broken Bond seeking revenge on Blofeld for killing his wife, but the story of Bond's marriage doesn't surface until the next film, *On Her Majesty's Secret Service*. Rather, *You Only Live Twice* was a hodgepodge of plot ideas hatched by the production team while on a long location scout in Japan. Newly minted megastar Sean Connery was at the center of the Bondmania that had gripped the globe, and Japan was no exception. Huge crowds followed the film crews in Japan to see the movie's shooting. One Japanese fan was so keen on Connery that they began following him with a camera, and the local police were brought in several times to stop crowd invasions during filming.

But behind the scenes, Connery was beginning to sow the seeds of discontent, and had started talking about leaving the franchise. The production team knew they needed to make the movie bigger than the actor playing Bond. To make the film a "happening," in those hippy days when "happenings" (performances, events, or situations meant to be considered art) were all the rage. It was time to go even bigger with James Bond, and a list of events and scenes was presented to none other than the now legendary children's author Roald Dahl. During WWII, Dahl and Ian Fleming had worked alongside one another as British officers promoting Britain's mission and message in the United States and combating the America First movement. Dahl had never written a proper screenplay before, so now it was time to pull off another literary feat—to conjure up some kind of story for the *You Only Live Twice* movie.

## THE ULTIMATE SPY-FI BOND?

What Dahl delivered was a spy-fi mishmash of a movie. The film's piecemeal creation comes through in a plot somewhat uneven in tone: from almost claustrophobically tense scenes, such as the poison dripping down the yarn toward a sleeping Bond, to scenes so grandiose and comic they feel like something from a superhero movie, such as the famous yellow gyrocopter, Blofeld's spectacular volcano base, and the helicopter with a magnet that carries off a carload of criminals chasing Bond. (When the criminals are dumped into the sea, Bond quips, "Just a drop in the ocean.") Then there's the downright daftness of having 007 trying to blend in with his Japanese surroundings, something Connery had signally failed to do during the real-life hysteria of Bondmania. All told, *You Only Live Twice* marked a brief pause in the idea of Bond as a serious spy series. And yet the movie's campiness and science fiction elements make it a livelier, more mercurial movie, the kind of film that would resurface later when Roger Moore played Bond.

Blofeld's cunning cosmic plot is to steal the space modules of both the US and USSR so that each superpower suspects the other. Then, when tensions have passed a nuclear peak and America and Russia have annihilated one another, a new superpower arises in the vacuum left—SPECTRE, of course. So the chase is on, and Bond is on Blofeld's tail. An array of science and tech is employed in Bond's mission, most of it associated with the gyrocopter. The actual flying machine seen in the movie was a miniature autogyro craft, the Wallis autogyro, developed in England in the 1960s by

British aviator and inventor Wing Commander Ken Wallis. Wallis's autogyros have been used in various ways, including military training, police reconnaissance, and even in a search for the Loch Ness Monster. Bond set designer Ken Adam had heard of Wallis on the radio and asked for a demonstration of the machine, dubbed Little Nellie by Wallis himself. According to Adam, Wallis flew for around forty-seven hours on the movie, and up to heights of ten thousand feet.

Machine guns, high explosives, two rocket launchers firing heat-seeking missiles (sixty a minute according to the screenplay!), flame guns, two smoke ejectors, and aerial mines. This is the tech we witness being used in the scene where Bond, camera on helmet for recon, battles against four normal choppers as he searches for Blofeld's volcanic base. The battle between Little Nellie and the helicopters proved rather tricky to get on film. The initial scenes were filmed in Miyazaki, Japan, with more than eighty-five takeoffs of the gyrocopter, and with Wallis almost crashing into the camera several times. Another scene, involving filming the helicopters from above, created such a huge downdraft that cameraman John Jordan had his foot severed by the craft's rotor. What's more, as the concluding shots involved explosions, the Japanese government didn't allow any further filming in a national park, so the crew moved to Torremolinos, Spain, which luckily resembled the Japanese landscape.

## BLOFELD'S VOLCANIC BASE
In a piece for *The Guardian* in 2002, Ken Adam explained the inspiration behind his spectacular set for *You Only Live Twice*:

> During the last week of location scouting in Japan, we flew over this incredible volcano, and we thought: wouldn't it be interesting if the villain was inside one of those? That appealed to me. I started scribbling. Cubby [Broccoli] and

the director, Lewis Gilbert, saw the scribbles and said, "Looks interesting, how much is it going to cost?" I had no idea. Cubby said, "Can you do it for a million dollars?" I said, "For a million dollars, I think I can." And then my troubles started. It's a mark of the franchise's success that, within five years, the production team had gone from a total budget of one million dollars for *Dr. No* to one million dollars being spent on a single set for *You Only Live Twice*.

Constructed under canvas at a lot inside Pinewood Studios, and with an operative heliport and monorail, the 148-foot-tall set could be seen from three miles away, and attracted many people from the area. Just in case you thought such bases were only found in fiction, look at this account of "Fortress London" from Mick Herron's book, *Slow Horses*: "In Central London, there's almost as much city beneath the streets as above. If the worst happens, whether toxic, nuclear, natural, or civil, these are the redoubts from which control will be reasserted. They are fundamental to London's geography, and appear on no A–Z."

## BLOFELD IS BUSTED

As the story of *You Only Live Twice* unfolds, we see Bond get the measure of Blofeld, but we also get a better measure of Bond himself. Connery's Bond continues in his vein of being confident without being smug. Sure, he's slick with the intel, yet Connery was never blessed with a whole lot of finesse. Barely two movies back, in *Goldfinger* he actually tried defusing the complex dirty atomic device by simply thumping at the thing with a gold brick.

When Bond gets to Blofeld's base, *You Only Live Twice* truly becomes a big spy-fi movie: it has scope and size that few films, even the Bonds that followed, have matched. We see Blofeld's huge rocket base hidden inside the colossal volcano set. We witness a helipad, a monorail, Mini Mokes, and an almost unlimited number of henchmen. And yes, the set was so huge they could, and did, fly a helicopter inside it. We see the superb helicopter shot of Bond fighting a dozen or so dockworkers as he runs across a warehouse roof, and we finally get to see Blofeld. Played by a very creepy-looking Donald Pleasence, Blofeld is a comic-strip villain. Looking remarkably like Dr. Evil, his parody in the *Austin Powers* film series, Blofeld has a huge vertical scar running through his right eye, and so once more is disfigured in the way of Doctors Rotwang, Strangelove, and No. As if to underline his villainy, Blofeld even uses a pool of piranhas with which he terminates underperforming employees, and insists on carrying a Persian cat that was so clearly traumatized it never worked in film again.

## MEANWHILE, IN THE SPACE RACE

But let's get back to that so-called space race between the US and the USSR; how did that race look during the filming of *You Only Live Twice* between July 1966 and March 1967? In February 1966, a Soviet space probe, Luna 9, had made the first controlled "soft" landing on the moon. The mission was celebrated as an engineering marvel, one which helped answer questions about the nature of the moon's surface and was to pave the way for the crewed missions that came later. Everyone expected those missions to be Soviet.

After all, Sputnik had started the space age. Humanity had seen Stone, Bronze, and Iron ages, and now the stars were our destination. The Soviets had already made an impressive start to the race to space: Sputnik had been Planet Earth's first artificial satellite; the Russian dog Laika had been the first living creature launched into space, also in 1957; and the first man in space, Yuri Gagarin in 1961, and the first woman, Valentina Tereshkova, in 1963, had followed in their wake. And in 1965, Alexei Leonov became the first human to walk in space.

## THE SOVIETS TRY ON BOND STYLE

The way in which the Soviets announced the success of their 1966 moon mission was worthy of a Bond plot. Rather than merely announcing their mission's February landing in the Ocean of Storms, the Soviets took a stealthier approach to global publicity. They simply sent back lunar pictures to Earth on a frequency that could easily be picked up. The Jodrell Bank radio telescope in England discovered them, to their surprise, as did many such stations across the world.

As well as sending back nine images, the Soviet mission answered a question that was seriously troubling both sides of the Iron Curtain. Many scientists feared that the lunar surface was shrouded in some sort of deep dusty "quicksand," and that any lunar lander would sink without trace. Luna 9 proved the ground was solid, a fact that helped the Russians and Americans move ahead with their manned programs. The director of the Soviet space program was a rocket engineer and spacecraft designer by the name of Sergei Korolev. After becoming frustrated with the endless and futile debate between Soviet scientists about the "hard" or "soft" nature of the lunar surface, he simply marched into his office one day and signed a decree that roughly read, "the lunar landscape is officially hard, signed Korolev." It's also rumored that the Nobel Prize committee tried to award Korolev but he turned the award down, as to award a single person for such a collective effort as the space program would be hubris beyond measure.

Piers Bizony, coauthor of *Starman*, a biography of Gagarin, suggests, "The Russians were in the business of conquering space . . . The Americans felt they were in a race and the nature of a race is that once you think you've won it you tend to stop running." As a result, many space historians feel that had the Soviets landed on the moon first, it is unlikely they would have abandoned it as swiftly as the Americans. The more authoritarian government of the USSR would have enabled them to spend money and marshal the talents of their population in a way that the US could not. Space historian Christopher Riley suggests that not only would the Soviet Union have continued with lunar missions, but they might also have built lunar bases.

So, at the time of filming *You Only Live Twice*, it really was possible that a communist nation would be the first to claim the moon for all mankind. To newspaper readers in 1966, it seemed the USSR was well on its way to beating the US in the space race, as indeed every milestone in the race to date had seen a Soviet victory. Eminent radio astronomer Sir Bernard Lovell, the director of Jodrell Bank, described the Soviet moon landing to the BBC as an historic moment and said it was the final achievement necessary for a manned landing on the moon. As Bond was chasing Blofeld about the globe, the future of the space race looked a lot more ominous for some than it later became. A spectre was still haunting Europe. The spectre of a Soviet-dominated space.

# ON HER MAJESTY'S SECRET SERVICE (1969)
# GUNS, GERMS, AND SUPER-VILLAINS

| *On Her Majesty's Secret Service* | **Year:** 1969 | **Bond:** George Lazenby | **Director:** Peter Hunt |
|---|---|---|---|
| **Producer:** Albert R. Broccoli & H. Saltzman | **Screenplay:** R. Maibaum & S. Raven | **Distributor:** United Artists | **Release Date:** December 18 |
| **Running Time:** 142 minutes | **Budget:** $7 million | **Box Office:** $82 million | **Body Count:** 42 |

## BOND AS SUPERMAN TAKES ITS TOLL

By now, James Bond was a spy-fi superhero. After the huge success of the Connery Bond films, the phenomenon of Bondmania, following in the footsteps of Beatlemania, gripped the culture for a five-year period in the mid Swinging Sixties. 007-related merchandise (especially kids' toys like the Lego, superhero, and *Star Wars*-type toys sold and marketed today), promotions, magazine articles, and television specials was everywhere. "Bond is important: this invincible superman that every man would like to copy, that every woman would like to conquer, this dream we all have of survival. And then one can't help liking him," said Sean Connery in an interview with Italian journalist Oriana Fallaci in 1965.

And yet, playing the superhero Bond character took its toll on Connery. Interviewing with Fallaci, published in her 1968 book *The Egotists: Sixteen Surprising Interviews*, Connery had identified the way in which he had molded Fleming's fictional character for the films:

I said to the producers that the character had one defect, there was no humor about him; to get him accepted, they'd have to let me play him tongue-in-cheek, so people could laugh. They agreed, and there you are: today Bond is accepted to such an extent that even philosophers take the trouble to analyze him, even intellectuals enjoy defending him or attacking him. And even while they're laughing at him, people take him terribly seriously . . . It became a terrible pressure, like living in a goldfish bowl . . . that was part of the reason I wanted to be finished with Bond. Also, I had become completely identified with it, and it became very wearing and very boring.

So, Connery took a break from Bond. After a hiatus of just one movie Connery returned to the role of 007 for *Diamonds Are Forever*. By that time, the head of United Artists himself had insisted Connery was to be seduced back to the franchise and that money was no object. Connery's fee? For resuming the role of James Bond he demanded, and got, £1.25 million (£26 million in 2018 money), or 12.5 percent of the movie's gross profits. But we're getting ahead of the plot. That Connery movie hiatus was *On Her Majesty's Secret Service*, in which Bond was played by Australian actor George Lazenby.

## BOND'S OFFER OF SPONSORED MARRIAGE

In *On Her Majesty's Secret Service*, Lazenby's Bond is confronted with a Blofeld whose plan is no less than to hold the world ransom with the threat of sterilizing Planet Earth's food supply. And how exactly would Blofeld achieve this? By using a bevy of brainwashed beauties known as the "angels of death," of course. How Bond and spy-fi is that? (By the way, the very look of the "angels of death" makes this one of the most dated-looking of all Bond movies, almost as if the filmmakers were trying to parody an *Austin Powers* movie.)

The tale proper begins in Portugal where Bond meets Marc-Ange Draco, the head of a European crime syndicate, Unione Corse. Draco tells 007 of the troubled past of his only daughter, Contessa Teresa "Tracy" di Vicenzo. Draco offers Bond one million pounds if he will marry her (this kind of thing only seems to happen to Bond). Bond refuses the generous offer, but agrees to continue wooing Tracy if Draco tells him the whereabouts of Ernst Stavro Blofeld, the head of SPECTRE. After a tip from Draco, Bond learns that Blofeld is corresponding with London College of Arms's genealogist Sir Hilary Bray. Blofeld seems to be trying to claim the title "Comte Balthazar de Bleuchamp." (Like Blofeld, the English seem to invest great importance in the question of ancestry. The entire British establishment became established simply by stealing land and property off everyone else, surrounding themselves with uncritical sycophants, and giving themselves titles with which they've wielded power ever since. No doubt Blofeld is after something similar.)

We see once more in *On Her Majesty's Secret Service* that Bond's facial recognition software seems to be acting up, as Blofeld is played by yet another actor, but since Bond himself has changed, we can assume that this Blofeld switch is an annoyance to the audience only. The three (visible) Blofelds are very different: Donald Pleasence was creepy, Charles Gray avuncular, and in this movie, Blofeld is played by a thuggish Telly Savalas. Savalas's Blofeld is a malevolent gangster. Like Bond, he's a man of action, heading up a ski chase after 007 escapes, and later brawling with Bond aboard a speeding bobsled. With his robust build and deep voice, Savalas is perhaps more convincing as a leader of men and a megalomaniac. He also commits the most ingenious action Blofeld ever performs on screen: he creates an avalanche by firing a flare into a mountain. Elsewhere, Blofeld seems to keep his alleged genius out of sight.

## POISON, FAMINE, AND PESTILENCE

Ancestry and Blofeld's true identity notwithstanding, Bond poses as genealogist Sir Hilary Bray. This is quite distracting as it means that not only is Bond dressed in traditional Scottish kilt for a good deal of the movie, but he also sounds like a nerdish snob, and a far cry from the subtle Scottish burr of Connery. Nonetheless, the kilted Bond travels to meet Blofeld. He's holed up in the Alps. The film location used was Piz Gloria, a revolving restaurant on the Schilthorn near Mürren in the Bernese Oberland, Switzerland. Here, Blofeld's diabolical master plan is hidden in a headquarters which claims to be a clinical allergy-research institute.

Bond meets twelve young women, patients at the institute's alleged clinic, who have been apparently cured of their allergies. Naturally, Bond, in his romantic element, goes on the prowl at night to the room of one patient, Ruby, in order to inevitably seduce her with his irresistible 007 charisma. But at midnight, Bond notices that the delectable dozen go into a sleep-induced hypnotic state, as Blofeld issues audio instructions (through the institute's PA system) for when they return home. The truth is revealed: the women are the "angels of death," the bevy of beauties being brainwashed to take bacteriological warfare agents out into the world.

In his 1909 short story "Yah! Yah! Yah!" American novelist Jack London wrote an account of a punitive European expedition to a South Pacific island, which purposely exposed the Polynesian population to measles, of which many of them died. (London was not exaggerating. History shows the British had deliberately tried to spread smallpox among North American Indians during Pontiac's Rebellion between 1763 and 1766. During a parley, the British gave Native representatives blankets and a handkerchief enclosed in small metal boxes that had been exposed to smallpox, in the hope of spreading the disease to the Natives and ending the siege. It's unclear whether the action was responsible for the estimated 400,000 to 500,000 Native Americans who died from smallpox during and after the war.) London had written a further sci-fi story the following year, "The Unparalleled Invasion," in which Western nations wipe out China with a biological attack.

## A BLOFELDIAN BLOT ON THE BRITISH LANDSCAPE

Secret documents suggest that the British Ministry of Defence (MoD) turned large parts of the UK into a giant laboratory to carry out a series of secret germ warfare tests on the British public. The official government report into Britain's biological weapons trials between 1940 and 1979 wasn't released until decades later, but they were clearly being conducted while *On Her Majesty's Secret Service* was filmed.

With a seemingly Blofeldian turn of mind, more than one hundred covert experiments were done, many involving the release of potentially toxic chemicals and microorganisms over huge swaths of the population, without the public being told. And with an almost Bond-villain-like level of bullshit, the government report reveals that military personnel were briefed to tell any "inquisitive inquirer" that the trials were part of research projects into air pollution and the good old British weather.

The tests, carried out by British government boffins at Porton Down in England (not far from Stonehenge), were done to help the MoD assess Britain's susceptibility to Russian attack. In many cases, it is claimed, the tests didn't use biological weapons but "harmless" alternatives, which boffins believed would mimic germ warfare. However, families in some areas whose children had birth defects demanded a public inquiry.

Take just three examples of these tests. First, in the "Sabotage Trials" between 1952 and 1964, bacteria was released to assess the vulnerability of large government buildings and public transport to attack. This included, in 1956, bacteria released on the London Underground at lunchtime along the Northern Line. Secondly, in tests between 1964 and 1973, germs were stuck to the threads of spiders' webs in boxes in London's West End, Southampton, and Swindon to test how the germs would survive in different environments. And thirdly, the infamous anthrax tests on Gruinard off the coast of Scotland, which left the island so contaminated it could not be inhabited until the late 1980s. Naturally, government spokespersons play down the entire affair, suggesting that eminent scientists have shown there was no danger to public health from these releases which were carried out to protect the public.

## THE ANGELS OF DEATH BRING FAMINE

Blofeld hatches a plan to use his "angels of death" to contaminate and ultimately sterilize the world's food supply according to their respective allergies—each an allergic reaction to staple foods from their homeland. Blofeld's plan is to simply rescind this threat if all his past crimes are pardoned and he is recognized as the current "Comte Balthazar de Bleuchamp."

Blofeld has done his homework. His Omega Virus threatens to wipe out entire species across the globe. Since the 1950s it has been possible to create deadly biological aerosols through the use of what's known as bursting bomblet technology. By the turn of the decade in which Blofeld finds himself, a B-47 bomber dispenser could infect over half the population of a sixteen-square-mile area with infectious disease.

So, rather than dealing in melodrama, Blofeld is dealing in real possibilities. In the *On Her Majesty's Secret Service* movie, Blofeld speculates to Bond that the Virus could be engineered to target humans. Science fiction in 1969 perhaps, but certainly not now. The production of such a virus to target humans rather than other animals is a rather straightforward affair today, one which would be within the biotech potential of many labs around the world.

## THE GREAT ESCAPE

But Bond stops all that, naturally. Enlisting the help of crime lord Draco, Blofeld's facility is destroyed, and Blofeld escapes the carnage alone in a bobsled. As Bond pursues him, we are treated to the delight of a bobsled chase, rather than a car chase, including the very Mario Kart moment when a grenade explodes under Bond's bobsled. The chase ends when Blofeld becomes trapped in a tree branch with an injured neck. Bond actually marries Tracy in Portugal, and we witness what must surely be the most dramatic ever ending to a Bond movie. As the happily married couple drive away in Bond's Aston Martin, Bond pulls over to the roadside to remove flowers from the car. Suddenly (spoiler!) another car seemingly comes out of nowhere in which a neck-braced Blofeld helps commit a drive-by shooting of Bond's car. Tracy is killed in the attack and Bond is devastated. Movie ends.

Devastated may also be the word to describe George Lazenby's exit from the franchise. Despite giving us a James Bond capable of vulnerability, a man who can show fear and is not immune to heartbreak, Lazenby never actually signed a contract. Negotiations dragged on during production, and Lazenby was convinced by his agent Ronan O'Rahilly that the world's most famous secret agent would be archaic in a liberated 1970s, so Lazenby left the role before the release of *On Her Majesty's Secret Service*. As Lazenby said in an interview on *The South Bank Show* in 2008:

> Ronan O'Rahilly was the man who created Radio Caroline. He took me under his wing. He was anti-establishment and he could hurt the establishment by taking me away from it. He said, these guys [Broccoli and Saltzman] are monsters, they're gonna use you and spit you out. I felt that Ronan knew what he was talking about. It was the hippie movement. LSD was out there. Ronan was opening my eyes. I was really under his spell. He told me, once you get typecast as Bond, you can't

get different parts. There was hardly any young person that didn't have long hair. And you can imagine how I felt, with short hair, trying to get laid. I looked like a cop, or a waiter. And people were peace, not war. And Bond was about war. Ronan convinced me [Bond] wasn't going to survive and I was basically speaking his mind. I was not the way they wanted their James Bond to be. I'd blown my shot at being a big famous movie star.

# DIAMONDS ARE FOREVER (1971)
# BOND'S BATTLE WITH WEAPON WIZARDS

| *Diamonds Are Forever* | **Year:** 1971 | **Bond:** Sean Connery | **Director:** Guy Hamilton |
|---|---|---|---|
| **Producer:** Albert R. Broccoli & H. Saltzman | **Screenplay:** R. Maibaum & T. Mankiewicz | **Distributor:** United Artists | **Release Date:** December 14 |
| **Running Time:** 120 minutes | **Budget:** $7.2 million | **Box Office:** $116 million | **Body Count:** 49 |

The first thing he noticed was that Las Vegas seemed to have invented a new school of functional architecture, 'The Gilded Mousetrap School,' he thought it might be called, whose main purpose was to channel the customer-mouse into the central gambling trap whether he wanted the cheese or not.

—Ian Fleming, *Diamonds Are Forever* (1956)

## REGENERATING BOND

Connery thought he was out. And yet, after their Lazenby hippy experience, the studio pulled Connery back in. As we mentioned earlier, the offer to rehire the actor's talents and try to salvage a stuttering franchise was a staggering sum of £1.25 million (or 12.5 percent of the movie's gross profits). As a further incentive, United Artists offered to film two movies of Connery's choice. (By the way, £1.25 million is the equivalent of £26 million today; it's worth noting that in 2017 *The Hollywood Reporter* estimated that A-list movie stars routinely make £12 million to £15 million for top roles in big-budget films. So Connery's fee was more than Vin Diesel was paid for 2017's *The Fate of the Furious*, for example, or Sandra Bullock was paid for 2013's *Gravity*, and roughly the same as the projected figure Daniel Craig will be paid for the imminent twenty-fifth Bond movie, which has a working title of *No Time to Die*.)

Clearly, the producers thought Connery was worth the coin. As Roger Moore pointed out in the 2012 book *Bond on Bond*, "Sean *was* Bond. He created Bond. He embodied Bond and because of Sean, Bond became an instantly recognizable character the world over—he was rough, tough, mean, and witty . . . he was a bloody good 007." And once more, the set for *Diamonds Are Forever* was on a spy-fi scale. Producer Cubby Broccoli suggested one scene based on a dream he'd had where he'd been reunited with his former employer, the infamous American billionaire Howard Hughes. In the dream, Broccoli realized that Hughes had been replaced by an imposter. When Broccoli mentioned the dream to screenwriter Richard Maibaum, Maibaum used it as the basis for the scene where Bond visits Las Vegas billionaire Willard Whyte, only to find that Whyte has been replaced by Blofeld. Who else?

## ON THE TRAIL OF BLOFELD

*Diamonds Are Forever* is another action-packed spy-fi Bond movie. We get the usual armory of high-tech gadgets, including a set of false fingerprints. We're treated to a high-speed chase in the city of Las Vegas, and Bond driving a moon buggy, then a desert three-wheeler trike in the Nevada desert. (As Bond escapes from W. Techtronics and runs across what looks like a "moonscape" set, we see one of the astronauts moving in slow motion, and who seems to try stopping Bond, who runs past him. The sequence is very tangential to the plot, and is assumed by many to be a jab at the puerile "the-moon-landing-was-fake" conspiracy theories.) Finally, we get one of the longest-running climaxes in a Bond film: a huge shootout consisting of frogmen jumping out of helicopters to blow up an oil rig while also doing the crucial job of knocking off dastardlier SPECTRE operatives. But, importantly for this book, *Diamonds Are Forever* was also the first film showcasing a plot in which a laser-like satellite-mounted weapon was proposed for global warfare.

Bond is still on the trail of Blofeld, of course, who's finally taken to plastic surgery to alter his identity and explain away the fact that he's now played by English actor Charles Gray. (If only the Bond franchise had gone one step further in the science fiction aspects of Bond. They could have embraced a kind of Doctor Who-like regeneration idea for characters played by a host of different actors. It would explain not only the regenerative sequence of James Bonds from Connery to Craig, but also the evolving identity of Blofeld and the changing face of the CIA's Felix Leiter. Or they could even have suggested the earthlier idea, picked out from the cunning plot of *Thunderball*, that the British secret service was using SPECTRE's facial surgery techniques to disguise their most important agent.)

Despite all this, Connery doesn't seem to want to be in the movie, even though he's being very well paid for the privilege. Connery ambles about after Blofeld like some aging heavyweight boxer, flirting with the franchise, his obvious toupee sometimes slipping. He fights for his life, bangs his elbows as he scraps with a hoodlum inside a cramped Amsterdam elevator, and is even duffed up by a pair of cocky female acrobats in the Nevada desert. Connery seems knackered, out of shape, and halfway through the out door. And yet the movie is fascinating. Legend has it that the very last scene Connery actually filmed as the official James Bond was the one at the crematorium. Bond is knocked senseless, dumped inside a coffin, and shoved in the direction of the flames and hell-fire!

## A DIAMOND RING

The spy-fi plot seems to drift in and out of reception, like a long-distance radio signal. Bond infiltrates a Las Vegas diamond-smuggling ring.

Following the trail of this diamond ring (ha!), Bond goes to the Whyte House, a casino-hotel owned by the reclusive billionaire Willard Whyte. Later, Bond enters the seeming destination of the diamonds: a research lab owned by Whyte, where a satellite is being built by Professor Metz, a laser specialist. With the help of Whyte, Bond raids the lab and uncovers Blofeld's cunning plan: to create a laser satellite using the diamonds, which by now has already been launched into space. Blofeld intends to use the satellite to destabilize the world by destroying nuclear weapons in China, the Soviet Union, and the United States, then holding the planet ransom and auctioning nuclear supremacy to the highest bidder. (But Bond being Bond, 007 is also captivated by the obligatory Bond girl, Tiffany Case, who may be a double agent.) "The satellite is at present over Kansas," Blofeld explains, eager to demonstrate the power of his outer-space warhead. "But if we destroy Kansas, the world may not hear about it for years."

## A WORD ABOUT SPACE LASERS

Once more, Blofeld is trying to use weapons as a form of blackmail. In the case of *Diamonds Are Forever,* we must surely wonder where Blofeld is getting this idea for a super-weapon spacecraft that can laser-beam to destruction missiles and armaments on the ground. It's probably too early for these ideas to have been inspired by nature itself, which is far more modest than Blofeld. Natural lasers, known as masers (**m**icrowave **a**mplification by **s**timulated **e**mission of **r**adiation), have been produced for billions of years in the atmospheres of stars, comets, star-forming regions, supernova remnants, and, allegedly, even super massive black holes. But during Blofeld's day, the universe had so far managed to keep quiet about its technological achievements.

Essentially, what happens with space masers is this: light and explosions from stars lend energy to nearby gas regions. These nearby gases, excited by collisions with the light, shift to higher energy levels in their structures. As with any laser, the "beaming" process has to start with a spontaneous emission of light accidentally shooting off in some direction. With space masers, the nearby light and explosions from stars leads to a continued stimulated emission, which means the maser beam heads off towards infinity; infinity is stretching the truth a bit, but it does sound rather good and would surely impress the likes of Blofeld.

## THE CULTURE OF SUPER-WEAPONS

As it's still too early in 1971 for Blofeld to have seen *Star Wars* and the Death Star in action, he must have delved deeper into the history of science fiction's obsession with super-weapons. Take H. G. Wells's *The War of the Worlds*, for example. The Martians come to Planet Earth with the intention of farming humans. Their super-weapon of choice? The death ray (Blofeld must surely have been impressed with the name). The death ray's beam annihilated everything it touched: matter into flame, water into steam, human flesh into vapor. If Blofeld did his research correctly, he would have found out that the discovery of radioactivity and X-rays in the late nineteenth century made for even madder ideas about super-weapons. In George Griffith's 1911 novel, *The Lord of Labour*, a war is fought with disintegrator rays, which triggered the use of some kind of "awesome" ray in many stories since, including Blofeld's. Percy Westerman's *The War of the Wireless Waves*, written in 1923, was a typical example, which features an arms race, in this case the race between the British ZZ rays (not merely a single Z, but a more deadly double Z) and the undoubted menace of the German Ultra-K ray (Ultra here doing the job of the deadly double Z of the British). In this culture of criminal scientists and super-villains who can weaponize their aims and missions with marvelous but deadly rays, or even atomic disintegrators, it should come as no surprise to us that a spy-fi villain like Blofeld sits so easily in a history of comic books and superheroes, which is also littered with the gadgets of superpower.

Despite this culture of super-weapons, Blofeld's super space weapon is preposterous. He's meant to have made a weapon powered by mirrors that collect, store, and focus the sun's light into a heat ray that is able to destroy the ICBMs of the superpowers. But, even allowing for the dubious idea that Blofeld actually knows where China, the USSR, and the US have hidden their weapons, the spacecraft's mirrors, which are made from diamonds for no obvious scientific reason, are far too small for the job. If you look at the size of the weapon's solar panels, the spacecraft is tens of meters in size. Too small to harvest enough solar energy to zap targets on the planet down below. How small, exactly? Well, the idea of harvesting sunlight and beaming it down to help generate electricity on Earth predates *Diamonds Are Forever* by thirty years, if you count Isaac Asimov's 1941 short story "Reason." In most schemes, including Japan's space agency, JAXA, the mirrors are kilometers in size. And by the time the beam reaches the earth's surface, it's even wider.

## DIAMONDS ARE FOREVER: THE MISSING LINK

Watching *Diamonds Are Forever* play out, it's easy to get the impression that it's not only Connery who seems like an aging heavyweight boxer. Bond must wonder if Western democracy is still worth saving. His mission feels fatally flawed from the start. Nobody seems to care. After coming across the moonscape set, 007 steals a car and screeches across the city with the Vegas cops in his wake. In the midst of this chase, film director Guy Hamilton suddenly, mischievously, decides to change our perspective. He cuts away to show the chase from the lugubrious interior of a Las Vegas arcade. In this gambling den, under glass like exhibited creatures of scientific interest, we spy the tourists lined up like zombies at the slot-machines, totally oblivious to the screeching tires and wailing sirens merely yards away from them.

Each era gets the Bond it deserves. Cubby Broccoli's franchise began self-assuredly, armed with a sleek moral sense; a secret agent hunting an absolute truth in a world of clearly defined good and evil. But by the mid-to-late seventies, the franchise had slipped into a jokey self-parody. *Diamonds Are Forever* stands on the cusp. The movie's a kind of missing link, the vital transitional Bond film in which the old glamour has lost its sparkle and the resolute hero has lost his way. For many, it's at times a blackly comic movie in which a beleaguered Bond does his best for a beleaguered Western world.

# LIVE AND LET DIE (1973)
# FULL THROTTLE: BOND AND THE CAR

| *Live and Let Die* | **Year:** 1973 | **Bond:** Roger Moore | **Director:** Guy Hamilton |
|---|---|---|---|
| **Producer:** Albert R. Broccoli & H. Saltzman | **Screenplay:** Tom Mankiewicz | **Distributor:** United Artists | **Release Date:** June 27 |
| **Running Time:** 121 minutes | **Budget:** $7 million | **Box Office:** $161.8 million | **Body Count:** 13 |

Women are often meticulous and safe drivers, but they are very seldom first-class. In general, Bond regarded them as a mild hazard and he always gave them plenty of road and was ready for the unpredictable. Four women in a car he regarded as the highest potential danger, and two women nearly as lethal. Women together cannot keep silent in a car, and when women talk, they have to look into each other's faces. An exchange of words is not enough. They have to see the other person's expression, perhaps to read behind the others' words or analyze the reaction to their own. So two women in the front seat of a car constantly distract each other's attention from the road ahead and four women are more than doubly dangerous for the driver not only has to hear and see what her companion is saying but also, for women are like that, what the two behind are talking about.

—Ian Fleming, *Thunderball* (1961)

## AN EYEBROW-RAISING INTERNATIONAL PLAYBOY

The financial success of *Diamonds Are Forever* wasn't up to the level of *Thunderball* or even *Goldfinger*. But it still beat *You Only Live Twice* and *On Her Majesty's Secret Service* at the box office. Broccoli and Saltzman tried to seduce Connery into making another movie, but the definitive Bond declined. So the producers turned to Roger Moore, an actor they'd considered for *On Her Majesty's Secret Service*, but who had been unavailable due to his television commitments as Robin Hood-like criminal Simon Templar in *The Saint* and Lord Brett Sinclair in an action and adventure comedy series called *The Persuaders!*

In both of these television roles, Moore was cast as a charming and debonair international playboy. So when Moore said he didn't want to try imitating Connery's Bond in *Live and Let Die*, screenwriter Tom Mankiewicz fitted the screenplay around Moore's screen persona. Bond became more comedic, with a lighter-hearted feel. It was an approach that led Raymond Benson, American author of the official Bond novels from 1997 to 2003, to describe Moore's Bond as "a rather smarmy, eyebrow-raising international playboy who never seemed to get hurt."

## BLAXPLOITATION

*Live and Let Die* was filmed at the height of the blaxploitation era of the early 1970s, which was influenced in turn by the Black Power Movement of the late '60s, and especially the early '70s. The movement sought racial pride, economic empowerment, and the creation of political and cultural institutions for African American people in the United States. Black Power activists founded bookstores and schools, farms and food co-ops, media and printing presses, clinics and ambulance services. The cultural impact of the movement spread to the Black Power Revolution in Trinidad and Tobago in the Caribbean. Perhaps the most relevantly successful blaxploitation movie for Bond was the famous 1971 movie *Shaft*, which brought the black experience to the silver screen in a revolutionary new way. It showed black social and political issues, which had previously been ignored in cinema.

The blaxploitation movies got some stick for stereotypical characters. And yet they were among the first to portray black people and communities as heroes and subjects of the films, rather than just sidekicks or victims of violence. Blaxploitation films were also among the first to feature movie soundtracks of funk and soul music. (It's worth noting that Paul McCartney's wonderful theme song for *Live and Let Die*, nominated for an Academy Award and perhaps the best theme of the entire franchise, also features a reggae break in the middle of the song.)

Indeed, a number of blaxploitation archetypes appear in *Live and Let Die*, including derogatory racial epithets and black gangsters. To some extent, the film departs from former Bond plots about megalomaniac super-villains. It focuses instead on drug trafficking, a common theme of blaxploitation movies. *Live and Let Die* is also set in African American cultural centers such as the Caribbean, New Orleans, and Harlem. (Injecting blaxploitation themes into Bond was not without significant problems. Black actor Yaphet Kotto, who played a starring role as the villain in *Live and Let Die* and also starred as a ship's engineer in the 1979 sci-fi movie *Alien*, later stated, "There were so many problems with that script . . . I had to dig deep in my soul and brain and come up with a level of reality that would offset the sea of stereotype crap that Tom Mankiewicz wrote that had nothing to do with the black experience or culture.") Filming started in October 1972. By December, part of the production was being filmed on location in Harlem. The producers were required to pay a local Harlem gang protection money to ensure the crew's safety. When the money dried up, some exterior shots were in fact done on Manhattan's Upper East Side. *Live and Let Die* was the first James Bond film to feature an African American girl romantically involved with Bond, Rosie Carver, and also showcased the blaxploitation archetype of the pimpmobile.

## THE CARS THAT MAKE BOND BOND

This obsession with pimpmobiles is interesting, as Bond has his own infatuation with the car. By now, Bond movies had their own archetypes, familiar plot elements, and distinctive themes that had developed through the years. The flirting with Moneypenny; the assignment from M; the sci-fi sandbox sessions with Q; the meeting with allies like Felix Leiter; the girls, guns, and gadgets; the super-villains; the exotic international locations; the tongue-in-cheek humor; and, of course, the car chases. And if there's one single movie that best exemplifies Bond's obsession with the car chase, it's arguably *Live and Let Die*.

Unlike the Bond movies that come before or after, there's no one single car that stands out in *Live and Let Die*. In *Dr. No*, Bond drives a light blue Sunbeam Alpine convertible, and in *From Russia with Love*, he drives a 1935 3.5-liter Bentley drophead coupé Park Ward. Fleming had an obsession with Bentleys. In three of the 007 novels, *Casino Royale*, *Live and Let Die*, and *Moonraker*, Fleming has Bond drive a 1930 Blower Bentley. In *Casino Royale* Fleming writes that Bond bought the car "almost new" in 1933, and declared the car to be one of the last Blower Bentleys ever built. So, by the time of the movie *From Russia with Love*, the Bentley was over thirty years old, and yet seemingly very Bond-appropriate for an international secret agent with immaculate taste.

It wasn't until *Goldfinger* that Q gave Bond the iconic Aston Martin DB5, and the role of the car flipped from mere form to function, and the idea of the car in spy films changed for good. It's a mark of the Bond influence that the huge success of the toy DB5 even beat the Batmobile to the punch. The Batmobile first appeared in *Detective Comics #27* in May 1939, where it was depicted as a rather dull-looking red car. It wasn't until 1966 and the *Batman* television show that toy companies, including Corgi, started mass manufacturing Batmobiles with the prominent bat motif and wing-shaped tailfins.

Bond's DB5 in the *Goldfinger* movie is rammed to the brim with gadgets and modifications. It matters little whether the sliding steel plate at the car rear is impractical, as it would need to be so thin it wouldn't be able to stop a bullet. Or that the ejector seat would be better suited to a much larger and more robust military vehicle. The point of science fiction and spy-fi is not to be taken scientifically literally.

## BOND AND THE "WHAT IF" IDEA

In fact, the idea of the car in Bond movies gives us an opportunity to delve into this matter a bit deeper. Like science fiction, the Bondian sub-genre of spy-fi has a similarity to mainstream science. We can think of spy-fi as an imaginative device for doing a kind of theoretical science: the exploration of imagined super-villains and scenarios, of projected plots and devices in espionage. Scientists build models of hypothetical worlds and then test their theories. Spy-fi writers and moviemakers do something similar: they also explore hypothetical worlds. But they have more scope. Scientists are meant to stay within bounded laws. Spy-fi has no such boundaries. But we can see that a spirit of "what if?" is common to both science and spy-fi.

What if we could travel by jetpack, like in *Thunderball?* What if both the United States and Russia had their space modules stolen by a mysterious rogue agency, as in *You Only Live Twice?* What if a huge and powerful corporation tried to reboot the entire human race, like in *Moonraker?* Or, as in various movies in the Bond franchise, what if a car could swim, vanish, or fly? These are the ways in which Bond and spy-fi is an influence on modern science and culture. So it makes sense if a spy-fi movie wants to push the envelope on cars. What if a spy car had an ejector seat? What if, as well as a bulletproof body, the spy car also had an early prototype of the soon-to-be-common satnav? What if the car was kitted out with a tire shredder and machine gun functions and a set of revolving license plates?

## BOND AND THE CONCEPT CAR

After all, just think of the concept car. Concept cars, also known as prototypes, are cars made to showcase new concepts in design and/or new tech. General Motors designer Harley Earl is usually credited with inventing the concept car. He did much to popularize it through its traveling Motorama shows during the 1950s, when Ian Fleming was writing the first of the Bond novels. Concept cars are most often shown at motor shows, with their new and radical designs, but they are also increasingly featured in films, such as the Lexus 2054 in 2002's *Minority Report*, or the Audi RSQ in 2004's *I, Robot*, or the Mercedes AMG Vision Gran Turismo in 2017's *Justice League*. Yes, more often than not concept cars appear in sci-fi films. But some of the most famous concept cars have appeared in Bond spy-fi films.

Let's just take two concept cars from the rich history of Bond movies. First, the 1967 movie *You Only Live Twice* features the Toyota 2000GT Convertible. The Toyota 2000GT is a legendary sports car, often thought of as Japan's answer to the E-Type Jaguar. The Toyota is a sleek and swift two-seater, blessed with an elegantly long hood and pop-up headlights. It was, however, a little cramped for a Connery-sized Bond, so when the standard coupe arrived on set for the film, Connery simply didn't fit under the roof. Toyota was faced with a dilemma. Either hack off the roof so the car was cool enough for a Bond film, or completely redesign and reengineer a brand new 2000GT convertible in only two weeks. Toyota chose the latter. Their soft-top never made it into production. And if you fancy the regular coupe version today, you can expect to pay over $1 million.

Second, there's the Aston Martin DB10 in the 2015 movie *Spectre*. If you've ever wondered why Aston Martin replaced the DB9 with the DB11, the answer is the James Bond bespoke concept car the Aston Martin DB10. Created especially for *Spectre*, the DB10 became an instant favorite when it hit the screens, and is often voted best movie concept car. In the film's plot, the DB10 is meant to be driven by 009. But 007 "borrows" the car and floors it to Italy for one of the best-ever Bond chases, as the stunning DB10 is pursued by the very capable Jaguar C-X75 through Rome, only to meet a soggy end at the bottom of the Tiber.

## LIVE AND LET CHASE

*Diamonds Are Forever* raised the bar on the Bond car chase. Bond led the Las Vegas police on an epic chase in Tiffany Case's Ford Mustang, pulling off a few fancy tricks, including the famous two-wheel stunt through a narrow alley to make his swift escape (the chase also featured the moon buggy and a three-wheeler trike in the Nevada desert). In *Live and Let Die*, Bond closes in on the powerful Kananga, also known as Mr. Big and played by Yaphet Kotto. Kananga is running a globally threatening scheme using huge supplies of his manufactured heroin. As he tries to unravel Mr. Big's plan, Bond meets Solitaire, a beautiful tarot card reader whose mysterious gifts are vital to the crime lord's success. And so the stage is set for pulse-pounding action sequences involving voodoo, hungry crocodiles, and turbo-charged chases.

*Live and Let Die* gives us not just regular car chases—we also get a double-decker bus chase, in which the bus is decapitated by a low bridge. The chase involving the double-decker bus was cunningly done. The production crew bought a secondhand London bus and adapted it by having the top section removed. They then placed the top section back, but now running on ball bearings to allow it to slide off on impact with the bridge. The bus driver came from

London, too; the bus stunts were performed by a London Transport bus driving instructor named Maurice Patchett.

(Another iconic stunt in *Live and Let Die* played dice with the zoological unpredictability of the natural world: the crocodile-hopping scene. Here, as you probably know, Bond realizes his best escape route is to use crocodile backs as stepping stones. Film director Guy Hamilton used real reptiles for the stunt. They came from a crocodile farm in Jamaica's Montego Bay whose owner, Ross Katanga, was also invited to actually perform the stunt. Ross had inherited the farm when his own father was eaten by one of the farm's 1,300 crocs. It was Katanga who suggested the crocodile-hopping stunt. The scene took five takes to be completed, including one in which the last crocodile snapped at Kananga's heel, tearing his trousers.)

## ROCK THE BOAT

Then, of course, there was the now iconic speedboat chase. Rather ominously for some, the shooting of that iconic boat chase began on Friday the 13th in October 1972, though some days saw interruption due to flooding. The sequence sees Bond evade Kananga's men by stealing a speedboat and heading out into the Louisiana Irish Bayou, a body of water within the boundaries of New Orleans, but separated from it by undeveloped wetlands. In total, the production crew used twenty-six boats in the sequence, leaping over roadways and gliding over lawns. By shoot completion, the crew had totaled seventeen boats. To enable director Guy Hamilton to capture close-ups of Bond behind the wheel, Roger Moore had to learn how to drive the boats. The lessons came at a cost as Moore drove one of his rides into a boathouse and fractured his tooth in the accident.

The sequence's pièce de résistance was the boat-jump at Crawdad Bridge over Sheriff J. W. Pepper and his car (it appears to be a coincidence that the Sheriff has the same name as Paul

McCartney's most famous album as a Beatle, by the way). To test if the stunt was safe, a mock-up of the levee and the police car was created by the crew, even down to a Sheriff J. W. Pepper made of bamboo. The plan was this: if stuntman Jerry Comeaux could perform the stunt three times without getting injured, then maybe it would be safe to put an actor in there! A crowd of locals and crew gathered to watch the stunt. They were not disappointed. On the second take, Comeaux's boat launched into a forty-foot leap, crashed into the water, skidded in the wake of another boat, and finally flipped up onto the bank. Thankfully, no one was hurt, but the sheer drama and impact of the stunt surprised even Comeaux. "When I finally saw film of the leap it scared the hell out of me." Indeed, the speedboat jump scene unintentionally set a Guinness World Record at the time with 110 feet cleared.

## SWIM, VANISH, OR FLY

To drive our point home (see what I did there?), let's look at three perfect examples that showcase the "what if" way that Bond spy-fi movies use the concept car; *what if a car could swim, vanish, or fly?*

### Swim

The 1977 Bond movie *The Spy Who Loved Me* features a Lotus Esprit S1 sports concept car that can swim. The Esprit was nicknamed "Wet Nellie" in reference to Little Nellie, the autogyro in *You Only Live Twice*. The custom-built machine was created specifically for the film by Perry Oceanographic, Inc., of Riviera Beach, Florida. They used a Lotus Esprit S1 bodyshell, which cost around $100,000 at the time. However, in the filming of the movie, the car didn't actually maintain a dry interior. It was a "wet sub" and its occupants needed to don scuba gear.

But here's the thing. The history of the machine after the movie is a good example of spy fiction into possible science fact. When filming was done, ultimately the sub went into storage. After the end of a ten-year lease, no one claimed the contents of the storage locker, so the sub was placed on auction. The buyer paid less than $100 for the unit, exhibited the sub occasionally between 1989 and 2013, then restored and placed the sub at auction as a Bond car in 2013. In September 2013 at RM Auctions in Battersea, London, the sub sold for £550,000. The machine was bought by Elon Musk, who plans to convert the car into the functional machine from the movie.

## Vanish

In the 2002 film *Die Another Day*, Bond is equipped with a new car: the Aston Martin V12 Vanquish, which is playfully nicknamed the "Vanish." The car is fitted with the usual Bond spec, but with one major new refinement: its notorious cloaking device. This allowed the Vanish to do just that; disappear and become practically invisible at the push of a button. The Vanish was heavily featured in the movie, especially in the Iceland scenes, where the car is involved in a chase sequence with a Jaguar XKR.

Of course, invisibility has been a sci-fi dream for generations. But the science of the matter may be finally showing promise, even with larger objects such as vehicles. You won't be surprised to hear that military applications head up the research, with Bae Systems having already demonstrated an infrared cloaking device for tanks. But we may never find a way to make a car like the Vanish become invisible to visible light. And, as you will read later in the *Die Another Day* chapter, the suggestion of an invisible car in the near future was not without its problems in the movie!

## Fly

In 1974's movie, *The Man with the Golden Gun*, Bond villain Francisco Scaramanga drives and flies an AMC Matador coupe, made by the American Motors Corporation from 1974 to 1978. The actual machine with strap-on wings that we see in the movie was not airworthy and could only make a 1,640-foot flight. But that's not the point.

As early as 1928, Henry Ford had realized the concept of the flying car in an experimental single-seat airplane he called a "sky flivver." According to a 2017 article in the London *Financial Times*, Ford had predicted in 1940, "Mark my words, a combination aeroplane and motor car is coming. You may smile, but it will come." Yet the first attempts with Ford's "sky flivver" were troubled; a pilot died in an early test flight.

And that's always been the problem with the flying car. It's technically doable, of course, but tricky in practice. Nonetheless, the role of Bond movies and spy-fi is to promote the very "what if" possibility of flying cars. It wasn't until February 2018, almost half a century after *The Man with the Golden Gun*, that Chinese drone manufacturer Ehang had their Ehang 184 craft fly forty journalists and officials on journeys of up to fifteen kilometers in Guangzhou, southern China. The 184 is said to have reached top speeds of around 80 mph, and to be robust enough to weather thunderstorms and typhoons. Though, in the event of any problems, the craft's system can revert to remote pilot from a control center.

Finally, at the beginning of this chapter on *Live and Let Die*, we quoted an example of Ian Fleming's now outdated sexism on the question of female drivers. As we have seen here, and will continue to see in further chapters and movies, it's Bond who is the "mild hazard" on the road, a man who should be regarded with "the highest potential danger"!

# THE MAN WITH THE GOLDEN GUN (1974)
# NOTES FROM A DYING PLANET

| *The Man with the Golden Gun* | Year: 1974 | Bond: Roger Moore | Director: Guy Hamilton |
|---|---|---|---|
| **Producer:** Albert R. Broccoli & H. Saltzman | **Screenplay:** R. Maibaum & T. Mankiewicz | **Distributor:** United Artists | **Release Date:** December 19 |
| **Running Time:** 125 minutes | **Budget:** $7 million | **Box Office:** $97.6 million | **Body Count:** 6 |

We stand now where two roads diverge. But unlike the roads in Robert Frost's familiar poem, they are not equally fair. The road we have long been traveling is deceptively easy, a smooth superhighway on which we progress with great speed, but at its end lies disaster. The other fork of the road—the one less traveled by—offers our last, our only chance to reach a destination that assures the preservation of the earth.

—Rachel Carson, *Silent Spring* (1962)

## KUNG FU AND THE OIL SHOCK

Roger Moore's second appearance as Bond is seen by some as one of the most disappointing of the 1970s 007 movies. *The Man with the Golden Gun* was originally planned as a follow-up to *On Her Majesty's Secret Service*, but it was hurried into production soon after *Live and Let Die* opened in an attempt to alleviate Harry Saltzman's non-movie money worries. In spite of the veteran hand of Bond director Guy Hamilton, who'd worked on *Goldfinger*, *Diamonds Are Forever*, and *Live and Let Die*, *The Man with the*

*Golden Gun's* script continues the tendency to flirt too far with ill-advised forays into cheesy comedy. These forays include the return of redneck US Sheriff J. W. Pepper, as seen in *Live and Let Die*, and diminutive supporting villain Nick-Nack. The movie also tries to jump on two contemporary and fashionable bandwagons; the first was the global craze for kung fu (the scene in a martial arts school was no doubt inspired by the success of Bruce Lee's 1973 movie, *Enter the Dragon*; so too perhaps was the decision to film much of the movie in Southeast Asia), and the second was the energy crisis.

Filming for *The Man with the Golden Gun* began on November 6, 1973. Just a month before, the 1973 oil crisis had begun. Members of the Organization of Arab Petroleum Exporting Countries (which included Kuwait, Libya, Saudi Arabia, Algeria, Iraq, Syria, and Egypt) had proclaimed an oil embargo. The embargo was a by-product of the Yom Kippur War, and was targeted at those nations thought to be supporting Israel. At first, those targeted nations were the United States, the United Kingdom, Japan, Canada, and the Netherlands. But the embargo was later extended to Portugal, Rhodesia, and South Africa. By the time production on the movie had been wrapped in Pinewood in August 1974, the price of oil had already risen from $3 per barrel to nearly $12 globally, with prices in the United States much higher. It was the embargo that caused the so-called oil crisis, which had many short- and long-term effects on world politics and the global economy.

## SINGLE-SHOT SOLUTIONS

This "first oil shock," as it was later dubbed, also led to far more interest in renewable energy, nuclear power, and other fossil fuels. As if still reeling from the crisis, and the realization of vulnerability to the Arab states, Western energy policies since 1973 have been dominated by crisis-mentality thinking. Expensive quick fixes such as war and invasion have resulted, along with single-shot solutions, rather than basic long-term research allowing plenty of scope for entrepreneurship and innovation.

One such single-shot solution is *The Man with the Golden Gun*'s "solex agitator." In a loose adaptation of Fleming's novel, the movie has M send Bond after this solex agitator, a device that can harness the power of the sun. M also tells Bond to square off against the assassin Francisco Scaramanga, the titular "man with the golden gun." (By the way, this name of Scaramanga is so clearly the most Bond villain name imaginable that even my word processing software wants to change it into Scaremonger!) In short, Bond has to find Scaramanga before Scaramanga kills Bond. To give our villain an extra diabolical edge, we hear that not only has he killed other 00 agents in the past, including 002 in Beirut in 1969, but he's also been trained by the deadly KGB (shock horror).

Like many Bond villains, the compelling Scaramanga hits the baddie benchmark. He is, of course, played by the wonderful British actor, Christopher Lee. Lee was not only related to Ian Fleming—they were step-cousins—but Lee was also one of Fleming's first choices for the Bond role. This choice was, in part, based on Lee's WWII record and his experiences in Winston Churchill's elite Special Operations Executive, whose missions remain classified to this day. They reportedly included conducting espionage, sabotage, and reconnaissance in occupied Europe against the Axis powers. As a result, Scaramanga compares very well with the host of wonderful Bond villains, which includes not just the more obvious Connery-era super-villain choices such as Blofeld and Goldfinger, but also more nuanced characters such as Elliot Carver and Le Chiffre.

## TWO OTHER PIECES OF TECH

### The Golden Gun
Apart from the solex agitator, the two pieces of standout tech in the movie are the golden gun itself and a flying car. First, the golden

gun. As props, three golden guns were made. They comprised one solid piece, one used in action that could be fired with a cap, and one that could be assembled and disassembled (though Christopher Lee said that the process was "extremely difficult"). For many, the golden gun is one of the more memorable props in the Bond series. The golden gun also ranked sixth in a 20th Century Fox 2008 vote for the most popular film weapons, a poll that surveyed around two thousand film fans.

The gun's tech consists of an interlocking fountain pen (which is the gun's barrel), a cigarette lighter (the gun's bullet chamber), a cigarette case (the handle of the gun), and a cufflink (the gun's trigger). The bullets were kept in Scaramanga's belt buckle. In the movie, the golden gun was designed to use a single 4.2-millimeter, 23-carat gold bullet produced by Lazar, a Macau gunsmith whose specialty was custom weapons and ammunition. Coincidentally, on October 10 of the same year as that 20th Century Fox poll, one of the golden guns used in the film was missing (suspected stolen) from Elstree Props. At the time, the gun's estimated value was around £80,000.

## The "Flying" Car

We've already mentioned the flying car in this movie, but there's also the film's famous corkscrew jump of an AMC Hornet over a twisted, broken bridge in Thailand. After a tense chase through the streets of Bangkok, Bond and his cheesy comic relief partner, Sheriff J. W. Pepper (who just happened to be in the passenger seat during the car's theft), end up on the opposite side of a river from Scaramanga. Bond lines up the AMC Hornet with a collapsed bridge, guns the engine, speeds across the bridge, takes flight across the water (with a barrel roll in the process), and lands in one piece at the other side of the river. All in one take.

(American racing driver Jay Milligan had invented the stunt. He had performed it in 1972 at the Houston Astrodome in an

AMC Javelin, rather grandly calling the stunt "The Astro Spiral Jump." Milligan was wise enough to get in touch with the Bond producers, possibly having seen car stunts in earlier Bond movies. The producers snapped up the stunt, promptly patenting the jump and preventing it from appearing in any other subsequent films.)

Spy-fi and engineering have always enjoyed a strong mutual cross-fertilization, and this was the first ever movie stunt that was extensively computer-modeled. Setting up a complicated stunt such as this is not easy. The production team needed to rehearse the jump in the controlled environment of a stunt show, and then carry it out for real on a movie set location in Thailand. In an attempt to prevent the jump from murdering any of their stunt drivers in action, the team set about engineering the problem, making sure the test crashes happened in the controlled environment rather than into a Thai river. To help with this engineering problem, the team turned to one Raymond R. McHenry. McHenry was working with Calspan Corp, which had started out as the Cornell Aeronautical Laboratory. Their specialty was research into instrumenting aircraft to learn about the underlying physics of flight. But the company had also begun logging data on the definitive physics of how cars moved, and possibly crashed, through the air. To help calculate their findings, they needed to make a model that would reliably test the data without constantly crashing cars, so a mathematical computer model was made called the HVOSM (Highway Vehicle Object Simulation Model). Now, it's very possible that "Highway Vehicle Object" is the nerdiest ever reference to a car, but the HVOSM got Bond across that bridge in one piece!

## BACK TO THE POWER BEHIND THE PLOT

In the movie's narrative, Scaramanga is essentially Bond's dark mirror image, and on a personal level the plot is about the two men locked in a fatal duel for supremacy. Scaramanga, it seems, is a hitman who needs only one shot. He's obsessed with the seemingly invincible 007, the only man worthy of being his equal. So Scaramanga carries out a hit on Gibson, Britain's missing solar-energy expert whose solex is so efficient (95 percent, thank you very much) that it could solve the energy crisis in one fell swoop. No wonder Scaramanga is keen. He has his dwarf henchman Nick-Nack steal the solex and take it to Scaramanga's secret island (there's often a secret island in Bond films).

007 then travels to Bangkok to meet Hai Fat, a wealthy Thai entrepreneur whom Bond suspects of arranging Gibson's assassination. But Hai Fat is later killed by that damned Scaramanga, who replaces Fat as the new chairman of the board. Bond is brave enough to fly a seaplane into dastardly (Red) Chinese waters, under the Chinese radar (presumably a worthy feat), and lands at Scaramanga's crib. On arrival, as is the civilized custom in the franchise, Bond is welcomed in a gentlemanly way by Scaramanga, who feeds Bond a civilized meal before he tries to shoot him. 007 is also shown around yet another villain's lair. The man with the golden gun's solar power plant is so high-tech that it's operated by a single henchman.

Scaramanga's aim is a monopoly on solar energy, or seducing the oil companies into buying him out. He intends to sell the tech to the highest bidder (these villains are so materialistic). The exchange between Bond and Scaramanga runs like this:

> **Scaramanga:** This is the collection point . . . I'm arranging for countries that can afford the price to send experts here to see this.
> **Bond:** But no solex until the money is in the bank, right?
> **Scaramanga:** Right.
> **Bond:** I have run across similar situations.
> **Scaramanga:** Not what I've got here. This way the highest bidder can build hundreds of stations and sell franchises for

hundreds more. He will literally have the sun in his pocket. A monopoly on solar power.

And in a response that directly references the 1973 oil crisis, Bond replies: "The oil sheikhs will pay you just to keep solar energy off the market."

It then seems that the screenplay writers have a crisis of confidence. Perhaps, as this is Bond, plain old solar energy might be too unexciting for fans of 007. Or maybe moviegoers will think that the solex agitator is the dullest MacGuffin in cinema history. So they have Scaramanga build an associated super-weapon. He decides to dramatically demonstrate the inherent potency of his kit by pointing a powerful solar beam at Bond's plane, and we see it instantly ignite. The film's finale features a duel between Scaramanga and Bond, which settles the fate of the solex.

## SET THE CONTROLS FOR THE HEART OF THE SUN

Of all the renewable energies Scaramanga could pick, solar is the best and most dramatic choice. Even at a distance of over ninety-three million miles, seen through the mist on a crisp, foggy morning, someone as scientifically illiterate as Scaramanga would know that, up close, this big ball of dim light is a fiery and energetic monster. (If you doubt my judgement on Scaramanga's scientific abilities, check out the scene with him and 007 at the solex; Scaramanga seems to know the cost of everything and the science of nothing; he appears to know less about his own tech than even Bond!)

Without sunshine, there'd be no heat, no food, no weather, no light, no day or night, and no seasons. In fact, no Earth as we know it. The sun is a huge furnace, almost one million miles across, a huge fireball that uses nuclear energy to light up the solar system. The very constituent atoms of hydrogen gas at the sun's core fuse together to make new atoms, releasing huge amounts of energy. The sun is so massively huge that the temperature at its very center is about sixteen million degrees centigrade, which is hot enough to fuse the hydrogen into helium gas at a rate of four million tons of gas every second. That's as much energy as seven trillion nuclear explosions, every second.

Should anyone imagine the solex agitator is the most fantastic single-shot solution associated with the sun, think again. That accolade goes to the Dyson Sphere. Compared to a Dyson Sphere, the solex agitator is a mere wristwatch. The idea of the Sphere is to construct a hypothetical megastructure that completely encompasses the sun and collects almost all of that warm sun energy for human civilization. The idea goes back to a paper, *Search for Artificial Stellar Sources of Infra-Red Radiation*, written in 1959 for *Nature* by British-American physicist Freeman Dyson.

The original idea was for a swarm of orbiting structures to capture a greater percentage of the energy radiating from a star like the sun. But Dyson soon realized that, inevitably, any growing civilization, especially one that might colonize the solar system, would need so much energy that its best option would be to totally surround the sun and harness the power of our star directly. The Dyson Sphere idea has appeared many times in popular culture. For example, a bona fide Sphere appears in the episode "Relics" of the much-loved TV show, *Star Trek: The Next Generation*. *Avengers: Infinity War*, the Marvel Cinematic Universe movie, depicts one of the Nine Realms—Nidavellir—as a Dyson Sphere, one which draws power out of a neutron star to satisfy its profound energy needs. The halos that feature in popular video games are conceptually mere strips or slices of Dyson Sphere in virtual form.

We shall return to strange structures in space with both *Moonraker* and *Die Another Day*, but for now we say goodbye to *The Man with the Golden Gun*. Bond naturally wins the duel, and the solex, just before the secret island is destroyed. He escapes unharmed with this movie's "love" interest, Goodnight, in Scaramanga's partly wind-powered Chinese junk. A fitting end to the franchise's flirtation with renewable energy.

# THE SPY WHO LOVED ME (1977)
# A NEW ATLANTIS

| The Spy Who Loved Me | Year: 1977 | Bond: Roger Moore | Director: Lewis Gilbert |
|---|---|---|---|
| **Producer:** Albert R. Broccoli | **Screenplay:** R. Maibaum & C. Wood | **Distributor:** United Artists | **Release Date:** July 7 |
| **Running Time:** 125 minutes | **Budget:** $14 million | **Box Office:** $185.4 million | **Body Count:** 147 |

The sea is everything. It covers seven-tenths of the terrestrial globe. Its breath is pure and life-giving. It is an immense desert place where man is never lonely, for he senses the weaving of Creation on every hand. It is the physical embodiment of a supernatural existence . . . For the sea is itself nothing but love and emotion. It is the Living Infinite, as one of your poets has said. Nature manifests herself in it, with her three kingdoms: mineral, vegetable, and animal. The ocean is the vast reservoir of Nature.

—Jules Verne, *Twenty Thousand Leagues Under the Sea* (1870)

## GOD SAVE JAMES BOND

By now, Roger Moore had truly become Bond. Several of Moore's personal penchants had been introduced into his portrayal of 007, including his preference for Cuban cigars and his questionable affection for safari suits (Moore's use of cigars contrasted with the cigarette-smoking Bond portrayals of Connery, Lazenby, and Dalton). Moore's Bond had come to represent a rather dated character, again in contrast to Connery, whose fashion sense had been a striking part of his portrayal in the 1960s. One movie critic, Andrew Spicer, has suggested Moore to be the most elegant and mannerly of the Bonds, with the persona of a refined English country gentleman:

> Roger Moore recreated Bond as an old-style debonair hero, more polished and sophisticated than Connery's incarnation, using the mocking insouciance he had perfected in his role as Simon Templar . . . Moore's humor was a throwaway, and certainly in the later films, verged on self-parody. It was an essential strand in the increasingly tongue-in-cheek direction of the series, which became more lighthearted, knowing, and playfully intertextual.

The self-parody that Moore brought to the role (once more in contrast to the more sober Bonds of Connery, Lazenby, and Dalton) was somewhat at odds with the hyper-absurd spy-fi settings in which Bond found himself, so Moore's Bond in *The Spy Who Loved Me* seems weirdly fitting in his social context. Britain in 1977 was at odds with itself: the nagging but anachronistic patriotism of the Queen's Silver Jubilee celebrations and the working-class rebellion that was punk. The Sex Pistols' revolutionary anthem, "God Save the Queen," was an attack on the Queen and the monarchy. The song got to number two on the UK singles charts that summer, though many believe it was actually the biggest-selling single in Britain at the time, but was kept off the number one slot for fear of causing offense. You can see why. The song included lyrics such as "God save the queen, the fascist regime/They made you a moron, potential H bomb/God save the queen, she ain't no human being/And there's no future in England's dreaming/Oh God save history, God save your mad parade/Oh Lord God have mercy, all crimes are paid." It was a Britain fractured, a country ill at ease with itself.

## TURBO-CHARGED *THUNDERBALL*

And so, in a subtle way, *The Spy Who Loved Me* exemplified Britain's conflicted internal battles. Moore's smirking Bond, at ease with his own lameness, a not-to-be-taken-seriously Dad-like figure adrift in a sea of punk, was the perfect Bond at the perfect time, and almost completely opposite to the spirit of Fleming's snobby patriotism. The original story was a slim novella, hardly featuring 007 at all, a tale told from the perspective of a Canadian motel employee with whom Bond has a brisk and loveless one-night stand.

Instead, *The Spy Who Loved Me* movie is turbo-*Thunderball*, from Barbara Bach's drop-dead looks to the submarine harpoon melee. The film has a classic Blofeld-esque Bond villain in Stromberg, whose world-shattering plan is to turn US and Soviet nuclear subs on their respective countries. The story starts with British and Soviet ballistic-missile submarines mysteriously disappearing. Bond is sent to investigate, as is Russian spy Major Anya Amasova, KGB agent Triple X, played by Barbara Bach. Microfilm plans for a highly advanced submarine tracking system are being sold in Egypt. Could this tracking system have been used to steal the subs? Bond and Amasova journey across Egypt to find out, where they come across Jaws, the infamously huge Bond assassin with razor-sharp steel teeth.

British-Russian competition between Bond and Amasova becomes warmer during the movie. Bond saves Amasova from Jaws, and their cooling rivalry turns to affection (all Bond movies, after all, need a love interest). The British-Soviet spy team identifies shipping tycoon and scientist Karl Stromberg as the man behind the weapon thefts. Posing as an (unlikely) marine biologist (no beard and far too snappily dressed, though thankfully no safari suit) and his wife, our dynamic duo visit Stromberg's base and find that he has launched a mysterious new supertanker, the *Liparus*, nine months earlier.

(On leaving Stromberg's base, by the way, they escape underwater in Bond's Lotus Esprit from Q Branch, which converts into a submarine. The Esprit is a superb spy-fi creation. It's the most gadget-ladened car since Q gifted Bond the legendary Aston Martin DB5 in *Goldfinger*. The Esprit has cement dispensers; a vertical missile rack that will knock a helicopter out of the sky [and does!]; front-loaded missiles; all the necessary functions to make the car's cabin watertight; further weaponry including mine dispensers, oil slick camouflage, and front loaded harpoons; and, of course, full submarine controls.)

## ATLANTIS

Our dynamic duo find that Stromberg is operating the stolen tracking system from his underwater *Atlantis* base. Soon, Stromberg's plan is revealed: the simultaneous launching of nuclear missiles from British and Soviet submarines to annihilate Moscow and New York. Stromberg wants to trigger a world nuclear war, one which Stromberg would survive in *Atlantis*. A new civilization would emerge, not on the obliterated land, but in the oceans, underwater. In Stromberg's words, "Why do we seek to conquer space, when seven-tenths of our universe remains to be explored? The world beneath the sea."

The technical challenge of making *Atlantis* a reality on the movie screen was led once more by production designer Ken Adam. As Adam told *The South Bank Show* in 2008:

> The films became bigger, and bigger budgets too. So, with *The Spy Who Loved Me* I decided to build a stage, the 007 stage. And then put the set inside it. And when you see the film again [Adam chuckles], and you see Barbara Bach being rescued by Roger, you had an enormous amount of water pumped into this corridor. And Barbara went completely hysterical, I've never seen her so frightened in my life. And that comes out in the film.

Adam also explained the more challenging aspects of the build:

> At first we thought of using an actual supertanker for the ship that swallows submarines—Cubby had some billionaire friend who was willing to let us have one—but we thought, there are a lot of fire effects, we would never get insurance. Cubby said, "So build a stage." A supertanker is divided into compartments, but to swallow up these submarines we needed a big space. I did it all out of steel and stainless steel cladding with two parallel elevators at the end. The set was opened by Sir Harold Wilson [former British prime minister]. Then there was Stromberg's apartment. I always tried to make the interior fit the exterior to some extent, but here I had a spherical exterior and a linear set. Nobody questioned how it was that you had a 100-foot palatial room inside that structure.

## LIVING LIKE STROMBERG

Stromberg's plan has a point, of course. The deep ocean remains one of the most mysterious places on the planet. We know less about the deep than we do about the surface of Mars. The Mariana Trench, those long, narrow chasms at the ocean's bottom, have an association with the movies. The deepest point of the trench is Challenger Deep. The immense pressure at this depth makes exploration very difficult; only four descents have ever been made. One of those descents was made by movie director James Cameron. While being a far cry from Stromberg's *Atlantis*, Cameron's submersible was nonetheless the first to take samples and photograph the sea floor.

And yet, the facts of the sea lead us to an inevitable questioning conclusion about Stromberg's plan: as the deep is as cold as ice, blacker than is imaginable, and with crushing pressures, how would Stromberg's new civilization cope in one of the most hostile places on the planet? Sure, *Atlantis* doesn't have to dwell in the deep, but how ocean-going is his idea? Real-life large and advanced underwater habitats were once more common than they are today. Of more than a dozen that once existed, only three remain, all based in the Florida Keys. One is the Marine Lab, which is used as a training and research facility by agencies such as the US Navy and NASA. The second is the Jules Undersea Lodge, which also offers undersea weddings and romantic breaks for up to a thousand dollars a night (it's unclear whether this vacation idea was influenced by *The Spy Who Loved Me*). Advocates of underwater habitation say it's not just research or even Bond-like romantic getaways that benefit. They suggest they could also help with overpopulation problems, or protect against the possibility of natural or Stromberg-like disasters that make land-based human life impossible. But how feasible is the life Strombergian?

## SURVIVING AT SEA: THE LIFE STROMBERGIAN

The technology to create underwater colonies of up to a hundred people already exists. Beyond that number, you'd need the kind of money Stromberg has to develop the technological advances needed to deal with emergency evacuation systems, and more sophisticated controls of humidity and air supply. Underwater habitats must run smoothly by monitoring life support systems—humidity, air composition, and temperature—from the surface. The third of the underwater habitats in the Florida Keys is the National Oceanic and Atmospheric Administration's Aquarius Reef Base. The base, which accommodates up to six aquanauts at a time, consists of a bright yellow circular disc tethered to the undersea lab that lurks sixty feet below. The disc collects data from a number of different sensors and sends it to shore using a special wireless internet connection. But future habitats in a more Stromberg mold could use satellites to relay this vital data. For Stromberg, energy independence would also be a challenge. Though placed at sea, it's clear that sustainable energy options, such as wave and solar power, are a very real possibility up on the surface.

What are the prospects of a larger civilization, sci-fi style, in the manner of the Gungans who live in underwater cities on the planet Naboo in *Star Wars*? Well, the Gungans are smart. Rather than creating one giant bubble, they make larger habitats by linking together a number of smaller bubbles, like a sheet of bubble wrap on the ocean floor. And that's the way Stromberg could go in Earth's oceans: multiple modules made of steel and glass. Such smaller structures would prove a very flexible living system. Modules could be added or taken away to create living space for as many people as Stromberg would allow. However, it's unlikely he would choose to build any deeper than one thousand feet, as the pressures at such depths, as James Cameron knows, would mean very thick walls were needed, along with very long periods of decompression for those citizens needing to travel to the surface.

Of course, long term, it's also possible that Strombergian aquanauts might evolve over time. We might be more flexible to life underwater than we think. Deep in the island archipelagos along the west coast of Thailand, there are small tribes called the Moken people, also called sea-nomads. Remarkably, Moken children spend hours in the ocean every day. Just like seals and dolphins, the children have wonderful control of their eyes. They can make their pupils smaller and change their lens shape, which allows them to see with perfect clarity underwater. So, would Stromberg's aquanauts evolve to survive underwater? It's perfectly possible. There are plenty of other mammals that live underwater and come up for air every now and again.

(In *Galápagos*, a novel by the well-regarded American writer Kurt Vonnegut, a small band of mismatched humans are ship-wrecked in the Galápagos Islands after a financial crisis cripples the global economy. When a disease then makes all humans infertile, the Galápagos band become the only fertile humans left on the planet, the last specimens of humankind, and their descendants eventually evolve into a furry species resembling sea lions, though as usual you never know whether to take Kurt Vonnegut seriously!)

## TAKING STROMBERG SERIOUSLY

The makeup of the air needed to sustain Stromberg's citizens would depend on the depth of the habitat. Current habitats use compressors to keep on pushing fresh air into the habitat through tubes that run down from the surface. Carbon dioxide is removed with a chemical compound known as Sodasorb. But below certain depths, additional measures would be needed to ensure a healthy ratio of oxygen to other gases in air as the body needs different levels of various air components when at pressure. Long-term, Stromberg would need to consider becoming self-sustaining, with habitats that grow plants, using natural or artificial light, to generate a fresh supply of oxygen.

What about food? Fresh seafood would be reasonably easy to get for Stromberg's aquanauts. (An aquanaut, like astronaut but under water instead of in space, may be defined here as an undersea explorer, especially one who works or lives underwater for long periods of time. Stromberg obviously qualifies but so too might James Cameron.) The regular spearing of fish and eating of plankton would be a start. Initially they could also supplement their seafood diet with canned, preserved, and dehydrated foods they'd hoarded in preparation. However, cooking would likely be avoided because, as with airplanes, fumes seem much stronger in static air.

Some people are actually preparing for a life underwater as a means of preserving human life in the event of an apocalyptic catastrophe. In a kind of reverse Noah situation, these ocean pioneers would take to the water. One such pioneer is Philip Pauley, a British futurist and founder of the London-based company that designed a self-sustaining habitat known as Sub-Biosphere 2.

Pauley's creation includes circular structures that can be floated out to sea and then sunk, making a haven for fifty to a hundred "fortunate" souls. Like Stromberg, Pauley is an obsessive. Working in fiction as well as fact, he's looking for a publisher for the first of a planned sci-fi trilogy called *The Moral Order*, a dystopian future in which a hidden underwater world is discovered. Pauley also believes, like Stromberg, that underwater living is a logical solution to the problem of environmental collapse as it would be easier and cheaper to pull off than founding space colonies.

Curiously enough, founding space colonies and remaking mankind is the subject of our next Bond movie in *Moonraker*. But for now, we leave Bond and Agent Triple X adrift in their own sub-biosphere, the escape pod that appears at the end of *The Spy Who Loved Me*. Bond seems familiar with the controls and maybe that's because the pod is similar to the version seen in *Diamonds Are Forever*. We leave the happy couple in this rather opulent vessel, equipped with a gold-plated ice bucket, for heaven's sake, a wide range of alcohol, several books, and a quaint carriage clock. What more could you possibly need, stranded at sea? A scantily clad Major Amasova lurks ready and waiting to kill her counterpart. If only he wasn't so damned charming.

# PART II: 1979–2002

# MOONRAKER (1979)
# DIVINING THE MASTER RACE

| *Moonraker* | Year: 1979 | Bond: Roger Moore | Director: Lewis Gilbert |
|---|---|---|---|
| Producer: Albert R. Broccoli | Screenplay: Christopher Wood | Distributor: United Artists | Release Date: June 29 |
| Running Time: 126 minutes | Budget: $34 million | Box Office: $210.3 million | Body Count: 81 |

## THE FORCE AWAKENS MOONRAKER

Bond movies are known for their cool locations. For the franchise is not just about spy-fi plots, turbocharged car chases, or beautiful and unreasonably accommodating women who double up as martial-arts masters. There's also the sense of exotic place. That trademark motif of iconic sweeping shots of scenery and far-flung destinations that many moviegoers will never see. Other than in a Bond film, naturally. There's Caribbean splendor to be had at Ocho Rios, Jamaica, in *Dr. No*, for example, or Paradise Island and Nassau in *Thunderball*. And then there's Phang Nga Bay, Phuket, in *The Man with the Golden Gun*.

*Moonraker* goes one giant step beyond. Bond travels not only to Venice, and Sugarloaf Mountain in Rio de Janeiro, but also to outer space. For *Moonraker* arrived in the wake of *Star Wars*. This new franchise on the block not only rebooted the genre of sci-fi, but also sent it into the stratosphere. No wonder the Bond producers wanted a piece of the action: in its limited release on May 25, 1977, *Star Wars: Episode IV—A New Hope* made $1.5 million in selected theaters, an impact that was unheard of at the time. The movie went on to make more than $775 million worldwide. The Bond producers had originally intended to film *For Your Eyes Only*, but instead chose *Moonraker* to cash in on the craze for space.

So *Moonraker* arrived perfectly timed, sitting as it did between *A New Hope* and *The Empire Strikes Back,* which hit theaters a year after *Moonraker*. The movie's press release boasted that there had never before been a set as big as the one built to house the film's city in outer space. The space station, inspired of course by Kubrick's *2001: A Space Odyssey*, is another artistic masterpiece by Ken Adam. As American movie critic Roger Ebert put it:

> One moment among many: when the station's artificial gravity is shut off, it appears to us that fifty or a hundred men are floating in free-fall in the command area. It's hard enough to make one man appear to be floating, but *Moonraker* doesn't spare the cash. [The movie's budget of $34 million is roughly the same as the budgets of the last four Bond films combined; and the film's worldwide gross of $210 million is more than any other Bond movie for the next sixteen years.] The stars of this movie are Ken Adam, the art director, and Derek Meddings, in charge of special effects. In addition to the gigantic space station, they provide lots of little touches, like 007's gondola in Venice, which turns into a speedboat and then miraculously grows wheels. *Moonraker* is a movie by gadgeteers, for gadgeteers, about gadgeteers. Our age may be losing its faith in technology, but James Bond sure hasn't.

## MOONRAKER, THE SPACE SHUTTLE FRONTIER

And yet the actual plot is far grander in its spy-fi scope than mere gadgetry. The hellish plan is to destroy most of human life on the planet by launching fifty globes from outer space, which would dispense a deadly nerve gas down into the earth's atmosphere. One part of this plot is the conquest of space. The movie soon reveals that a Drax Industries Moonraker space shuttle on loan to Britain is somehow hijacked in midair. So M, naturally, assigns Bond to investigate. At the Drax Industries shuttle-manufacturing complex in California, Bond meets Hugo Drax, the company's oligarch and our resident Bond villain for *Moonraker*. Almost immediately, the plot thickens. Astronaut Dr. Holly Goodhead also meets Bond, who survives yet another assassination attempt, this time while he's trying out a centrifuge chamber. Drax has his personal pilot, Corinne Dufour (who had earlier helped Bond), killed by his pet dogs (an early hint that Drax may be a tad villainous).

(The space shuttle, of course, went on to dominate American space policy and space culture for many years after *Moonraker*. The shuttle appeared in many movies and TV shows through the decades, including a shuttle-like craft in *Buck Rogers in the 25th Century*, in *Armageddon*, in *Top Gear*, and in a classic episode of *The Simpsons*. But only in *Moonraker* did the space shuttle become an icon before it actually flew.)

The main premise of *Moonraker* is the exploration of space. The pre-credits sequence of the movie is superbly dramatic. We see the nighttime flight of a space shuttle being piggy-backed and then launching atop a shuttle carrier aircraft. The production team may have opted for this representation of a shuttle launch after seeing the real space shuttle being transported to and from launch and testing sites on top of a modified Boeing 747, which effectively acted as a giant air ferry. NASA had begun work on the *Columbia* space shuttle in 1975. The *Columbia* was named after the American sloop *Columbia Rediviva*, which between 1787 and 1793 had become the first American vessel to circumnavigate the globe. After years of research and development, NASA had the ambition to launch the *Columbia* shuttle in 1979. But it would be two years until the technical challenges were overcome, so the timing of *Moonraker* was impeccable.

## NERVOUS WORLD

The other part of Drax's diabolical plan is the nerve gas. Not only does Drax need the spacecraft to put "a necklace of death about the Earth," but he also needs a nerve gas capable of killing one hundred million people in its designated area. (Maybe Drax has been talking to Blofeld, whose plan for world domination in *On Her Majesty's Secret Service* revolved around the use of germ warfare.) Bond finds out about the nerve gas when he again encounters Goodhead, this time in Venice. Goodhead is snooping around a glass factory. Bond is also chased through the canals by Drax's henchmen. When a speedboat full of hoods start gunning him, Bond conveniently finds the gondola far more teched up than he'd imagined: the gondola has speedboat controls of its own. But the gadget obsession goes maybe one step too far when Bond flicks another switch to inflate the gondola's unlikely hovercraft skirts, which somehow enables it to climb a set of steps and hover over St. Mark's Square. Incidentally, if you look closely you may be able to spot the unlikely spectacle of the "double-taking pigeon," an editing trick that makes it look like a bird in St Mark's Square cannot believe its eyes as Bond's gondola transforms into a hovercraft. The motif is a recurring one in the movies of director John Glen; a startled pigeon that makes the actor, and the audience, jump. You may also spot the motif in his other Bond films, though look out for variations. A cat is used instead of a pigeon in *A View to a Kill*, and a monkey is used in *The Living Daylights*. All of Glen's Bond films also feature a character who dies by falling from a height, in a sequence often accompanied by the same "male scream" sound effect.

When Bond returns to the glass factory at night, he finds a secret biological laboratory (the push-button door entry into the nerve agent lab is the *Close Encounters* theme; yet another timely reference to contemporary cinema's craze for all things space). Glass vials are being made to hold a nerve gas that is deadly to humans, but harmless to animals. Soon, Bond finds evidence that Drax is moving his operation to Rio de Janeiro, so the action switches to Brazil. There, Bond survives an attack by Jaws on the Sugarloaf cable car. (With the vast productions that Bond movies entail—four studios, seven countries, and three continents—not everything goes smoothly. The Rio cable car filming nearly fell apart when the owner of the cable car company decided it would ruin his business and so physically tried to stop the filming.)

Bond later learns that the nerve gas toxin comes from a rare orchid indigenous to the Amazon jungle where Drax has yet another base. The wonderful exchange between them reveals the second part of Drax's plan:

> **Drax:** The curse of a civilization. Neither war nor pestilence wiped out the race who built the city lying around us. It was their reverence for this lovely flower.
> **Bond:** Because long-term exposure to its pollen caused sterility?
> **Drax:** Correct, Mr. Bond. As you have discovered, I have improved upon sterility. Those same seeds now yield death. Not, of course, to animals or plant life. One must preserve the balance of nature.

## BOND IN ORBIT

Finally, we get to outer space. In the Amazon jungle! Bond had witnessed four Moonrakers lifting off and, with Goodhead, they pose as pilots on Moonraker 6. The shuttle docks with Drax's space station, hidden from radar by a cloaking device. And what a space station it is. As we've already seen, the budget for *Moonraker* was huge (bigger than the first *six* Bond movies combined). But that didn't mean the producers were going to throw money away at the exorbitant rates that a certain special effects company wanted to charge for the *Moonraker* space sequences, so those particular *Moonraker* sequences were filmed in-house. Rather than use the new computer-controlled cameras, which were then becoming commonly used for special effects, Bond moviemakers used the somewhat dated technique of shooting a single special effects element. It worked like this. Assume we want to create a sequence of the space shuttle moving across the frame. Once done, we then have to stop the camera, rewind the film, then create and film another element to the shot. Naturally, the technique was prey to entropy and the risk of scratching or breaking the film, or maybe even accidentally exposing it. It could ruin many hours or days of work. Indeed, the section of film of the complex space battle sequence in *Moonraker*, which involved dozens of elements, was treated as if it was worth its weight in gold.

## ARTISTIC SPACE STATIONS

*Moonraker*'s special effects are pretty decent for the time. But the set design and model work, as you'd expect, was even better. NASA had cooperated extensively with the *Moonraker* production team. Visits to both Johnson Space Center and Ames in California had happened, though no NASA facilities actually appear in the movie (there are some exterior scenes at hangar facilities in Palmdale in the Mojave Desert). And yet Ken Adam seemed rather unimpressed by NASA:

> We did the space-station centrifuge for real and as a model. I spent some time at NASA and they were very proud to show me all their concepts for space stations, which haven't really altered since—they're just a series of big cylinders bolted together. I felt that wasn't interesting enough for a Bond film. So, I came up with the idea of a mobile, a very irregular mobile, with these various arms, corridors to satellites, at various angles.

(By the way, around the time *A New Hope* was released, a group of professors met for a period of ten weeks to brainstorm the design of space colonies. Their recommendation? Purpose-made colonies for orbit around planets or moons. The colony would be a wheel-like habitat one mile in diameter. Colonists would live in a tubular wheel structure, which would rotate to simulate gravity, and use mirrors for solar power. It's not exactly the Death Star, but it's a start. And it's not a million miles away from Ken Adam's concept.)

Ken Adam's space station is important for another reason. It's a timely reminder that what really counts in Bond is the spy-fi nature of villain and plot. Sure, Adam knows the science of space stations and artificial gravity, what works and what might not. But the literal science is not his main concern. He needs to bend the rules of mere science and invoke his artistic license to paint us a picture of living in space. A license to make art, rather than a license to kill.

That's just how spy-fi, and sci-fi, work. This point about the duty of filmmakers like Ken Adam is nicely stated in the movie of Carl Sagan's novel *Contact*. The most dramatic episode occurs when Jodie Foster's character, scientist Dr. Ellie Arroway, goes on a galactic space flight. She is confronted with the visual marvels to be seen at the center of our galaxy. The awe and wonder of the universe. She is lost for words in the face of such beauty and humbly suggests a poet may have been a better choice for such a journey. It takes an artist to best express the taste, the feel, and the human meaning of scientific exploration.

## DIABOLICAL DRAX

Soon Jaws captures Bond and Goodhead. Drax then reveals his plan to destroy human life on Earth by launching the fifty globes that dispense the nerve gas into the atmosphere. It's a curious decision by Drax. It's true that the movie follows the novel's theme of death dealt from the skies above. Whereas Fleming's book boasts a weapon of mass destruction in the nuclear-armed missile, the movie opts for these metal spheres that contain the nerve gas, and which are launched from low-Earth orbit. But why deal with the complex scientific challenge of aiming and firing the spheres from space when they could be deployed remotely and more effectively on the ground? After all, surely the cost of Drax's space exploration (multiple shuttles *and* a space station, for heaven's sake!) would require a fortune beyond what one corporation could possibly afford (remember that the cost of building and running the International Space Station is a joint project between no less than five participating space agencies: Roscosmos, ESA, JAXA, NASA, and CSA). Drax should have opted for the cheaper ground-based option and spent more time with Blofeld; maybe some of his "angels of death" were still knocking around.

Drax explains that he had transported several dozen genetically perfect young men and women ("a race of perfect physical specimens") of varying races to the space station in the shuttles. They would live there until Earth was safe again for human life; their descendants would be the seed for "a new super-race." ("Your offspring will return to Earth and shape it in their image . . . From their first day on Earth they will be able to look up and know that there is law and order in the heavens.")

Drax, played by French actor Michael Lonsdale, has little humor about him. Some have suggested that his lines are delivered with such obvious boredom that one almost expects Drax to start snoring in the middle of one of his sub-Nazi monologues. The contention is that, since he's about to murder billions of people, there'd be a bit more glee about him. Even Hannibal Lecter had a sense of humor (though Darth Vader didn't).

## GO FASTER, NASA

Naturally, Bond quashes the diabolical plan. Drax's forces are defeated by space marines and the station is destroyed (the production team had great fun blowing up the station by inventively blasting away at it with shotguns on a closed set). Meanwhile,

Bond shoots Drax and ejects him into space. Bond and Goodhead use Drax's laser-armed Moonraker 5 to destroy the three already-launched globes and return to Earth. (I must admit on watching the movie again I only counted two nerve agent capsules being dispersed in space—would the nerve agent still not fall to Earth? Perhaps the calculation is that the toxin would burn up on entry—but not the third!)

*Moonraker* failed in its original goal to be a movie released into cinemas at roughly the same time that NASA's space shuttle took flight. But there was no publicity on which the movie could ride, as the shuttle's launch date slipped out of 1979 and 1980 and into April 1981. Nonetheless, NASA no doubt benefited from the cinematic exposure of the shuttle before it actually flew, even if it was depicted in the service of a Bond villain, and with go-faster stripes painted on its side. The Bond franchise had flirted once more with the conquest of space, albeit with a genocidal twist.

# FOR YOUR EYES ONLY (1981)
# THE THREADS OF
# NUCLEAR PARANOIA

| For Your Eyes Only | Year: 1981 | Bond: Roger Moore | Director: John Glen |
|---|---|---|---|
| Producer: Albert R. Broccoli | Screenplay: R. Maibaum & M. G. Wilson | Distributor: United Artists | Release Date: June 24 |
| Running Time: 127 minutes | Budget: $28 million | Box Office: $195.3 million | Body Count: 54 |

## BIG MONEY BOND

Legendary Bond producer Cubby Broccoli once said that "For every dollar you give away, you'll get a hundred back." The production ecologies of the Bond films show that Cubby's confidence in the economic logic of the film industry was well-founded: Broccoli had spent money in order to make money. (To put it another way, and to quote Henry Ford, founder of the Ford Motor Company, "I'm not in the business of making cars; I'm in the business of making money.") As the Bond movies became more and more successful, their production costs rose in order to make each film bigger and more spectacular than its predecessor. In Broccoli's words:

> I honestly feel a responsibility toward all the Bond fans out there. I know they look forward to these pictures, and so I'm going to go on delivering them as long as I can. Also, it's a challenge, which I enjoy. We try to make each picture more exciting than the one before, to take Bond somewhere he hasn't already been.

The importance Broccoli attached to spectacle and high production values was ramped up further in the promotion and PR for the films themselves. Each new film claimed to be the "biggest" and/or the "best" so far. For example, the tagline for *Thunderball* was *Here Comes the Biggest Bond of All!*, for *On Her Majesty's Secret Service* it was the rather curiously and hippy tinted *Far Up! Far Out! Far More!*, for *The Spy Who Loved Me* it was the most nakedly obvious *It's the Biggest. It's the Best. It's BOND—and B-E-Y-O-N-D*, and for *Moonraker* it was *Where all the other Bonds end . . . this one begins.*

Throughout the 1970s this correlation between investment and box-office profits held well. Broccoli doubled the budget of *The Spy Who Loved Me* from the previous film and went on to make a return over twice the amount of *The Man with the Golden Gun*. Individual Bond movies never hit the fiscal heights of super blockbusters such as *Star Wars: Episode IV—A New Hope*, *E.T.: The Extra-Terrestrial*, or *Titanic*. But as we saw in our chapter on *Dr. No*, when adjusted for inflation, the Bond franchise is the most successful in the history of cinema. Cumulative box-office grosses of the Bond movies passed $2 billion with *The Living Daylights*, $3 billion following *Die Another Day,* and Bond became the first franchise to reach $10 billion grossing.

## THE BOND FORMULA

Over time, the franchise became self-aware. The Bond films are such a unique and distinctive brand that the term "Bondian" was coined. The term is not used merely by those talking about the style and content of the films. It's also used by the moviemakers themselves. After observing the making of *The Spy Who Loved Me* for an Open University study of media production, Janet Woollacott noted, "'Bondian' was the phrase used by [Cubby] Broccoli and other members of the production team to mean 'in the spirit of James Bond' . . . To a certain extent the term 'Bondian' was used to describe the Bond films, which were seen as a distinctive formula, a specific genre of film."

To make the most of the money invested and render the movie truly spy-fi and Bondian, it's clear that a conscious decision was made to make the movies "of the age." As Ken Adam said about his sets for *Dr. No* and onward, "What I felt at that time—we're talking about '61—was that I couldn't remember seeing a film that reflected the age we were living in." So the Bond franchise placed itself very firmly in the present. Cashing in on cultural context has been just as successful as Broccoli's dollar giveaway. Witness the vital "what-if" context of the Cuban Missile Crisis to *Dr. No*, the 1973 oil crisis that acted as a crucial backdrop to *The Man with the Golden Gun*, or the imminent shuttle stage of space exploration and the cultural revolution that was *Star Wars* as the settings for *Moonraker*.

The consensus has been that some of the later Bond movies, such as *For Your Eyes Only*, had lost the creative edge and energy that made the earlier films so exciting. For example, Ronald Bergan in his 1986 book *The United Artists Story* suggested, "What was fresh in the 60s and had a certain faded charm in the 70s began to look dated and mechanical in the 80s, despite the usual technical expertise." And yet to write off the Bond films of the 1980s is pure folly. Rather than seeing them as tired, it makes more sense to recognize the success of the Bondian production strategy. The Bond films have always sought to revitalize and modernize their spy-fi formula: of course, this fact is most obvious in the way Bond himself is recast, but it's also evident in the nature of their spy-fi conspiracy plots, which are reactions to the changing political and technological landscape.

## BOND IN A NUCLEAR CONTEXT

It's no surprise that with *For Your Eyes Only* and *Octopussy*, Bond went fully nuclear. As far back as 1955, with his *Moonraker* novel,

Fleming had posed the rhetorical question, "who today would be satisfied with such an unambitious piece of villainy as a nuclear bomb hitting London?" The atom bomb had been invented in fiction only four decades earlier. In his prophetic 1914 novel *The World Set Free*, British science fiction writer H. G. Wells had been the first to christen the term "atomic bomb." ". . . And these atomic bombs which science burst upon the world that night were strange, even to the men who used them," Wells wrote in his story. His timetable for the development of nuclear capability was unnervingly farsighted, though a little conservative on dates. In *The World Set Free*, it wasn't until the 1950s that a scientist uncovers atomic energy and realizes there's no going back from this momentous discovery. Nonetheless, the scientist feels "like an imbecile who has presented a box of loaded revolvers to a crèche." Wells's book also foresaw a world war in 1956, with an alliance of France, England, and America against Germany and Austria.

After reading *The World Set Free* in 1932, Hungarian physicist Leo Szilárd became the first scientist to seriously examine the nuclear physics behind the fiction. Soon after, the Manhattan Project was masterminded. It would ultimately engage around 130,000 employees and cost a total of $2 billion (over $20 billion in today's figures). The project bore the fruit of the detonation of three atomic weapons by the United States in 1945. The Manhattan Project's lead scientist, Robert Oppenheimer, mulled over the "atomic bomb" first realized in fiction. Oppenheimer spoke for a generation of scientists when he said, "In some sort of crude sense which no vulgarity, no humor, no overstatement can quite extinguish, the physicists have known sin; and this is a knowledge which they cannot lose."

In the atomic age that followed 1945, leading up to Fleming's *Moonraker* novel, the world lived with a mood of apocalyptic terror. Millions of people responded anxiously to the catastrophic visions of speculative fiction, and to the reality of the apocalyptic threat of the bomb that fiction had been so instrumental in developing. At this focal moment in history, many felt that speculative fiction was alone in its ability to project ways out of this predicament. It became the means by which a mass audience was confronted with the possibility of holocaust and mutually assured destruction. No other literature came close.

## LIVING UNDER THE MUSHROOM CLOUD

America had believed itself invincible in the aftermath of the war, so when Moscow entered the atomic age, Washington was stunned. As the 1950s wore on, the Cold War climate found its frosty reflection in film and fiction. There was plenty to report. McCarthyist anti-communist witch-hunts, the conspiratorial House of Un-American Activities Committee, and growing international tension between superpowers equipped with the bomb. A slew of popular science fiction B-movies captured the mood of the era. *The Day the Earth Stood Still* (1951) told the tale of an alien who visits Earth to warn global leaders that taking their quarrels into space will mean fatal costs. *Invaders from Mars* (1952), *War of the Worlds* (1953), and *It Came from Outer Space* (1953) all cranked up the idea of imminent nuclear attack. *Invasion of the Body Snatchers* (1956) was a satire of McCarthyist paranoia in the early days of this ever-chillier Cold War.

Then there was *Dr. Strangelove*, Stanley Kubrick's famous satire of nuclear blunder, which Bond legend Ken Adam had worked on. Kubrick had felt personally vulnerable after reading the 1958 Cold War thriller *Red Alert* by Welsh author Peter George. The book was a grim warning. Its caveat was the absurd ease with which nuclear apocalypse may be accidentally triggered. *Red Alert* was serious melodrama, and Peter George a rather somber ex-RAF navigator who had recently joined the Campaign for Nuclear Disarmament. Kubrick viewed George's idea of chance catastrophe as too farcical for drama. It was a killing joke. In Kubrick's words, "How the hell could the President ever tell the Russian Premier to shoot down American planes? Good Lord, it sounds ridiculous." So *Red Alert* repeated itself, first as tragedy, second as farce, and *Dr. Strangelove* was born.

## FOR YOUR EYES ONLY

The threat of nuclear blunder also sits at the heart of *For Your Eyes Only* and *Octopussy*. Broccoli and Saltzman had all but invented spy-fi, and the genre that Hollywood calls "event films." The famous Bond formula we spoke of above, which Bond popularized and others copied, usually included super-villains, stunts, and sex, along with all-important and outlandish spy-fi plots to destroy or rule the world. After the *Star Wars*-esque special effects in *Moonraker* and the comedy-romp of *The Spy Who Loved Me*, Bond gets real and murderous with *For Your Eyes Only*.

A British military spy ship (disguised as a fishing boat) is sunk off the coast of Albania, and the planet's superpowers start a feverish search for its valuable lost cargo: the powerful ATAC (Automatic Targeting Attack Communicator) system. The ATAC is used by the UK's Ministry of Defence to communicate with and coordinate the Royal Navy's fleet of Polaris submarines. The ATAC gives its bearer control over Britain's Polaris nuclear capability, and its possession by the wrong hands (obviously, Russians) raises the specter of the unambitious piece of villainy of a nuclear bomb hitting London, which Fleming spoke of in his 1955 *Moonraker* novel.

Bond is naturally sent to recover the ATAC device (which has shades of other MacGuffin Bond devices, such as the Lektor and the solex) before it finds its way into the wrong hands (again, Russians). But why had the Bond franchise greeted this new decade with such stark realism? And why was the early '80s so reminiscent of the nuclear paranoia of the 1950s? The explanation could be that, in the 1980s, the USSR had overtaken the West in terms of the size of its nuclear stockpiles. Or could it have more to do with the election of the West's chief warmongers in Ronald Reagan and Margaret Thatcher? Our focus on what we might call "nuclear Bond" continues in the next chapter on *Octopussy*.

# OCTOPUSSY (1983)
# POLITICAL SCIENCE: WHEN TWO TRIBES GO TO WAR

| *Octopussy* | **Year:** 1983 | **Bond:** Roger Moore | **Director:** John Glen |
|---|---|---|---|
| **Producer:** Albert R. Broccoli | **Screenplay:** R Maibaum, G. M. Fraser & M. G. Wilson | **Distributor:** United Artists | **Release Date:** June 6 |
| **Running Time:** 131 minutes | **Budget:** $27.5 million | **Box Office:** $187.5 million | **Body Count:** 58 |

## BOND AND NUCLEAR PARANOIA

If *For Your Eyes Only* was starkly realistic, *Octopussy* was Bond back to its spy-fi and surreal best. This is the movie where Bond goes for an Indian: the suave and sophisticated super-villain Kamal Khan. Khan is a gem smuggler, and 007 picks up Khan's scent upon discovering a murdered British agent with a Fabergé egg. Once more the spy-fi plot of a Bond movie pushes the willing suspension of disbelief to the limits, and then stops. In *Octopussy* we have a plot that spans four continents, and includes Bond trudging around New Delhi in a gorilla costume borrowed from a traveling circus; a double-headed super-villain in the form of not just the smug smuggler Khan, but also his goggle-eyed conspirator in Steven Berkoff's Soviet General Orlov; Russian nukes; a jewel heist; an alligator automaton; and Bond's Tarzan impersonation while swinging from a vine. Director John Glen seems to have shared Stanley Kubrick's belief that the end of the world through nuclear blunder isn't tragedy, it's farce.

And yet who could blame the producers for creating a little escapism in those days of nuclear paranoia, for many moviegoers in 1983 would be suffering from nuclear paranoia as if it were a nagging migraine. For example, if you were watching a news report about growing military tensions along some border; if a plane passing overhead sounded a little unusual; or when the local municipal bureaucracy decided unexpectedly to test the town's sirens. All cause for that paranoid nuclear alarm. Has there ever been a bigger, more mind-boggling paranoia than imminent nuclear war? It trumps even the slow creeping death of climate change. No wonder the Bond producers cashed in on the crisis.

## LET'S DROP THE BIG ONE; SEE WHAT HAPPENS

But how was the culture of this crisis created? The election of British Prime Minister Margaret Thatcher in 1979 and American President Ronald Reagan in 1981 hadn't helped. In 1972, American songwriter Randy Newman had satirized American foreign policy with "Political Science," a song on his *Sail Away* album. "They don't respect us/So let's surprise them/We'll drop the big one/And pulverize them/Asia's crowded/And Europe's too old/Africa's far too hot/And Canada's too cold/And South America stole our name/Let's drop the big one/There'll be no one left to blame us."

It's unclear whether President Richard Nixon was a Randy Newman fan, but halfway through his most controversial of presidencies between 1969 and 1974, Nixon had begun to adopt a fresh approach to international relations. Rather than seeing the world as a hostile and bipolar place of superpowers, he used diplomacy rather than military power to try to create a plurality of poles. Nixon encouraged the United Nations to recognize the communist Chinese administration. After a trip to China in 1972, Nixon initiated diplomatic relations with Beijing. He also helped thaw frosty relations with Russia, adopting a policy of détente toward the Soviet Union. And in 1972 he signed the Strategic Arms Limitation Treaty with Soviet premier Leonid Brezhnev, which prohibited the manufacture of nuclear missiles by both sides, taking a sensible step toward reducing the decades-old threat of nuclear war.

Despite Nixon's efforts, however, the Cold War got hot again under Reagan. When Reagan first entered office, you may remember, he is said to have asked for the way to the Pentagon War Room. Reagan was referring to the entirely fictional, though clearly impressive, control room, which Bond production designer Ken Adam had built for Stanley Kubrick's dark comic movie masterpiece, *Dr. Strangelove*.

Reagan was always going to be different from Nixon. As President of the Screen Actors Guild, he had appeared in 1947, along with Walt Disney, as a friendly witness for the House of Un-American Activities Committee. Reagan spied reds under the bed: he testified that a small clique within the union was adopting "communist-like tactics" in an attempt to influence union policy. Reagan had long believed the spread of communism threatened freedom. Consequently, Reagan's foreign policy (known as the Reagan Doctrine) was more of the Randy Newman kind, giving financial and military aid to anticommunist governments and insurgencies around the world. Based on the perspective of free-lance academic Rob Miller, it's easy to see that the Reagan Doctrine was a reversion to type. Miller has written:

> The US Government had four years to conceive and instigate [a] propaganda campaign beginning in 1941 when the USA entered WWII after the attack on Pearl Harbor. By the time of the dropping of the first atomic bomb on Hiroshima, ideas of fighting "the enemies of freedom" and the need for military superiority were already well established in the public psyche . . . in essence, the instigation of Cold War

propaganda campaigns was simply a matter of replacing Nazism with Communism as "the enemy of freedom."

## MWA-HA-HA-HA!

*Octopussy* tapped into the nervous world of the early 1980s. Bond is sent to solve the murder of fellow agent 009, who was killed in East Germany clutching a fake Fabergé egg. The trail soon leads to India, where an enigmatic woman operates an unlikely smuggling ring under the cover of a traveling circus. Naturally, as this is spy-fi, her real motives prove to be far more sinister, and Bond uncovers a plot to blow up a US Air Force base in West Germany.

And yet it's the portrayal of the Soviets that seems straight out of a spy-fi comic book. We kick off with a Soviet face-off between Generals Gogol and Orlov. General Gogol probably represents actual Soviet policy of the time, but it's General Orlov who gets all the screen time. He wants to send tanks rolling into Europe. It's hardly a subtle dissection of Soviet foreign policy in the late 1980s; at one point a Soviet representative screams, "the West is decadent and divided!" And on another occasion, someone says "world socialism will be achieved peaceably" without the slightest hint of irony. All the script is missing is the odd "mwa-ha-ha-ha" to complete the comic book characterization.

Traveling to East Germany, Bond becomes a clown in the circus and discovers that Orlov has swapped Soviet treasures for a nuclear warhead, which is primed to explode during the circus show at the US Air Force base. According to the simplistic plot, the explosion would trigger Europe into seeking disarmament in the belief that the bomb belonged to the US, and had detonated by mistake, leaving borders open to a Soviet invasion. Bond realizes Orlov's tactic toward the end of the film:

> My God. Of course. Our early-warning system will rule out the bomb having come from Russia or anywhere else.

Everyone will assume incorrectly that it was an American bomb triggered accidentally. Europe will insist on unilateral disarmament, leaving every border undefended for you to walk across at will. It doesn't matter a damn to you that thousands of innocent people will die in this accident of yours. On your feet, General. You're going to stop that train.

## BLUNDERS AND TRIBES

Finally, and within months of *Octopussy* being released, the world almost died. Stanislav Petrov, a Soviet lieutenant-colonel in the USSR secret service, eased himself into the commander's seat in the underground early warning bunker south of Moscow. As the flying time of an ICBM from the US to the USSR, and vice-versa, was around twelve minutes, if the Cold War should ever run hot, men like Petrov would have only seconds to make the call between life and death. And that's exactly what happened.

Warning lights in front of Petrov suddenly flashed up, and system sirens wailed. The message seemed clear: the US had just gone to war. The Soviet computer had detected missiles being fired. And yet the screens were showing nothing untoward; no tell-tale flash of a missile launching out of its silo into the sky. Armageddon, or system glitch? Instead of panicking and calling an alert to send Soviet missiles in retaliation, Petrov decided to wait. Even though the system now claimed further missiles were actually in the air ("Missile attack imminent!") Petrov stuck to his solid-sense reasoning. This had to be a mistake, as so few missiles were showing and not an expected mass attack. And so the Soviet alert on September 26, 1983, had been a false alarm. And Stanislav Petrov's cool head had saved the world.

The following year on June 4, 1984, "Two Tribes," an anti-war song by British band Frankie Goes to Hollywood, was released. Originally recorded for a BBC John Peel session in October 1982, after filming on *Octopussy* had begun, "Two Tribes" was a nihilistic, gleeful lyric expressing enthusiasm for nuclear war. The accompanying video featured a fictional wrestling match between Reagan and then-Soviet leader Konstantin Chernenko in front of an assembly of representatives from the world's nations; the wrestling match eventually degenerates into global destruction. The single was a phenomenal success in the UK. It immediately entered the charts at number one and stayed at the top for nine consecutive weeks. It was the longest-running number-one single in the UK of the 1980s. It wasn't just the Bond producers that cashed in on the nuclear crisis.

# A VIEW TO A KILL (1985)
# DEATH TO SILICON VALLEY!

| A View to a Kill | Year: 1985 | Bond: Roger Moore | Director: John Glen |
|---|---|---|---|
| Producer: Albert R. Broccoli & M. G. Wilson | Screenplay: R. Maibaum & M. G. Wilson | Distributor: United Artists | Release Date: May 22 |
| Running Time: 131 minutes | Budget: $30 million | Box Office: $152.4 million | Body Count: 62 |

**Zorin:** For centuries alchemists tried to make gold from base metals. Today we make microchips from silicon, which is common sand. But far better than gold. For several years we've had a profitable partnership: you as manufacturers, while I passed on to you industrial information that made you competitive, successful. We are now in the unique position to form an international cartel to control not only production but distribution of these microchips. There is one obstacle—Silicon Valley in San Francisco. [*After one member falls to his death*] Well, does anyone else want to drop out?

—Michael Wilson and Richard Maibaum screenplay,
*A View to a Kill* (1985)

## HE'S GOT TUSKS

And so, it's goodbye to Bond. Well, goodbye to Roger Moore's Bond, anyhow. With 1985's *A View to a Kill*, Moore agreed to appear in his seventh and final film; the same final number of Bond films as Sean Connery. For some, Moore, at the age of fifty-seven, was a little long in the tooth, especially since his female costar Tanya Roberts was only thirty. As *The Washington Post* said of Moore's age, "Moore isn't just long in the tooth—he's got tusks, and what looks like an eye job has given him the pie-eyed blankness of a zombie. He's not believable anymore in the action sequences, even less so in the romantic scenes." If that sounds a little harsh, then consider the fact that Moore himself later said he not only felt "a bit long in the tooth," but also said in 2007 that he "was only about four hundred years too old for the part."

And yet, by the time Moore retired in 1985, the oldest actor to play 007 in the series, his Bond movies had grossed over $1 billion at the box office. Moore's triumph in the role, especially in meeting the impossible mission of replacing the seemingly irreplaceable Sean Connery, is no doubt one of the main reasons the franchise has lasted. But by 1985, Moore was looking a few movies too old for swashbuckling and romancing. He no longer resembled his likeness in that opening gun barrel sequence. And one wonders, when watching the movie again, whether they're using the stuntman, rather than Moore, even for sequences such as climbing a flight of stairs or sipping a martini (it's not so easy to tell, as the actor's hair and that of his stunt double have been dyed the same color).

## 1984 WON'T BE LIKE *NINETEEN EIGHTY-FOUR*

By 1985's *A View to a Kill*, the computer revolution was well under way. In a now iconic and legendary commercial, Apple had launched itself into the world during Super Bowl XVIII on January 22, 1984. The build-up had been dramatic, and Apple's Macintosh computer the most anticipated personal computer release ever. Part and parcel of the promo was movie director Ridley Scott's "1984" commercial. Apple had commissioned Scott to direct a one-minute movie advertisement for the Mac, costing around three hundred thousand dollars. The advertisement was based partly on George Orwell's novel *Nineteen Eighty-Four*. And did Scott deliver.

The commercial opens on a dystopic scene, all dull blue and grayish tones. We see a line of androgynous minions marching as one through a tunnel, evoking the opening scenes of *Metropolis*. Rows of telescreens monitor their every move. In sharp contrast, a nameless runner appears, cast in full health and color. Dressed like an athlete, she is carrying a large brass-headed hammer. As the nameless athlete runs toward us, she is chased by four agents of the Thought Police. Dressed in black riot gear, and with visored helmets making them anonymous, the agents fail to keep up with the runner. She races toward a huge screen with the image of a Big Brother-like figure. In a version of *Nineteen Eighty-Four*'s "two minutes of hate," Big Brother dictates to the masses from a giant telescreen:

> Today, we celebrate the first glorious anniversary of the Information Purification Directives. We have created, for the first time in all history, a garden of pure ideology—where each worker may bloom, secure from the pests purveying contradictory truths. Our Unification of Thoughts is more powerful a weapon than any fleet or army on earth. We are one people, with one will, one resolve, one cause. Our enemies shall talk themselves to death, and we will bury them with their own confusion. We shall prevail!

Finally, as the dictator's rant is in full flow, the runner hurls the hammer at the giant screen, right at the moment Big Brother announces, "we shall prevail!" In a symphony of light and smoke, the huge image is destroyed, stunning the people transfixed by the telescreen. The people are freed. The commercial ends with a deep, portentous voiceover, as the scrolling typeface reads: "On January 24th, Apple Computer will introduce Macintosh. And you'll see why 1984 won't be like '1984.'" The screen fades to black as the voiceover ends, and the now iconic rainbow-Apple logo appears. The commercial, shown just once, was a major tipping point in the marketing of computers. And the whole idea for the advertisement was a veiled reference to IBM, Apple's main competitor.

## THE PARADISE OF LITTLE FAT MEN

In the history of sci-fi, of course, computers have often had a potentially problematic image. Many movies suggest that it'll only be a matter of time before computers become self-aware, rise up, and enslave or replace the humans who created them. Some of those silver-screen computers have been especially memorable, such as HAL 9000 in *2001: A Space Odyssey*; the Skynet AI in the *Terminator* franchise; and the Machine intelligence in the *Matrix* series.

Then there's George Orwell. Orwell's concern with progress and its social consequences led to his legendary book, *Nineteen Eighty-Four*. The novel was about the fears and frustrations of an individual caught up in a complex, overly rationalized society. It was a prophecy of a totalitarian future based on "not any particular country, but the implied aims of industrial civilization." Orwell was troubled by the devastating effects of science and technology. In his book, *The Road to Wigan Pier*, Orwell wrote:

> Barring wars and unforeseen disasters, the future is envisaged as an ever more rapid march of mechanical progress; machines to save work, machines to save thought, machines to save pain, hygiene, efficiency, organization . . . until finally you land up in the by now familiar Wellsian *Utopia*, aptly caricatured by Huxley in *Brave New World*, the paradise of little fat men.

Orwell foresaw the technology of control. The insidious nature of *Nineteen Eighty-Four*'s culture of surveillance stems from its telescreens and Thought Police. In Orwell's marvelous words, "The Beehive State is upon us, the individual will be stamped out of existence; the future is with the holiday camp, the doodlebug, and the secret police."

## A VIEW TO A KILL

The Bond producers, ever willing to make a movie fit for the times, chose the computer revolution as the topic of Moore's last outing as Bond. The plot of *A View to a Kill* revolves around a dictator of the spy-fi kind; a mad industrialist who plans to create a worldwide microchip monopoly by destroying California's Silicon Valley. The man with the plan is Max Zorin, played by the wonderful Christopher Walken. (The role of Zorin was also offered in turn to David Bowie, Mick Jagger, and Sting, but all three turned it down. Bowie said of the role, "Yes, I was offered that. I think for an actor it's probably an interesting thing to do. But I think that for somebody from rock it's more of a clown performance. And I didn't want to spend five months watching my double fall off mountains.") As the head of Zorin Industries, Max is bad. He's so bad, in fact, that not only was he born in Dresden, which became part of communist East Germany, but he is also ex-KGB (and also has a zeppelin)! And if that isn't bad enough, dear moviegoers, it's revealed later that Zorin is a steroid kid, the product of Nazi medical experiments, in which pregnant women were injected with huge quantities of steroids in an attempt to create "super-children." Most of these experimental pregnancies failed. But a few babies survived. They grew, like super-Max, to become not only extraordinarily intelligent, but also extraordinarily psychopathic, naturally—this is spy-fi.

Bond finds a microchip on the body of dead colleague 003 in Siberia. Rather late to the game, and clearly implying 007 has never read Orwell's *Nineteen Eighty-Four*, Bond then decides that *computer technology has the potential for sinister applications*! It's his investigation that leads Bond to discover Max Zorin and his less than subtle scheme to dominate the world. As head of Zorin Industries, does Zorin want to create armies of slave-droids under his command? Does he perhaps choose to create something even bigger and better than HAL, or maybe make his own version of Skynet (hell, he could even call his machine intelligence Zeus or something similar)? Nope, Zorin's diabolical plan for the future of computing is to take over the microchip industry by simply flooding Silicon Valley. And to help Zorin create a deluge in Northern California's global center for high tech, he has at his side the mighty and mysterious May Day, played by Grace Jones, a woman of very few words and ominous intent.

## FROM BUSINESS SOLUTIONS
## TO THE FINAL SOLUTION

As to Bond discovering that computer technology has the potential for sinister applications, that may have happened long before *A View to a Kill*. Recently found files from Hitler's Germany suggest that American computer company IBM supplied the Nazis with technology that was used to help transport millions of people to their deaths in the concentration camps at Auschwitz and Treblinka. That's the controversial claim of Holocaust expert Edwin Black in his book *IBM and the Holocaust*.

Black's book says there is documentary evidence that IBM executives at the company's New York base directly controlled a Polish subsidiary. It was this subsidiary that leased punch-card machines to "calculate exactly how many Jews should be emptied out of the ghettos each day" and transport them more efficiently on the busy railways, which led to the camps.

With books like Black's, the spy-fi of Bond doesn't seem so fantastic. And Black's hypothesis gets even harder-hitting. He maintains that when the Nazis invaded Poland, IBM New York set up this new subsidiary and the company's "sole purpose was to service the Nazi occupation during the rape of Poland." The subsidiary was named Watson Business Machines, after its then-president Thomas Watson, and they even operated a punch-card printing shop over the street from the Warsaw ghetto, according to Black.

As evidence for his spy-fi hypothesis, Black quotes one Leon Krzemieniecki, who was the sole surviving person employed by the Polish administration of the rail transport system to Auschwitz and Treblinka. According to Black, Krzemieniecki said that he "knew they were not German machines . . . The labels were in English . . . The person maintaining and repairing the machines spread the diagrams out sometimes. The language of the diagrams of those machines was only in English."

The book details the way in which payments for the lease of the Hollerith card machines were sent from Poland through Geneva to IBM in New York. A member of Mr. Black's research team, Robert Wolfe, who was formerly in charge of Nazi files in the US national archives, said the sheer detail of the evidence may silence would-be detractors. "The word has gotten out, and a lot of people still alive are supplying information that they didn't have the context to understand before. For those who have complained the proof is not there, this leaves little room for deniability."

The company, whose company headquarters are now based in Armonk in Upstate New York, has not denied the role of its subsidiaries in aiding the way the Nazis managed the Holocaust. In their defense, they suggested they should not be held responsible for the actions of companies of which the Third Reich had seized control.

It's unclear whether IBM's New York executives knew the Mephistophelian use to which their machines were being put. And yet Black details evidence of examples of stories in American newspapers at the time that he claims should have left IBM in little doubt about the nature of the Nazis' activities in Poland. Black also suggests that "IBM recovered all its Polish profits and machines" after the Nazi surrender. And a 2002 article in *The Guardian* went one step further. It said that Thomas J. Watson, CEO of IBM at the time, received a medal from Hitler for his services to the Third Reich. There is even a suggestion that Watson himself directly profited from the Holocaust, personally taking 1 percent of all IBM profits. IBM has named one of their current AI projects "Watson" in honor of him.

## THE GREATEST CATACLYSM IN HISTORY

Max Zorin's spy-fi plan to dominate the world relied on the machinations of geology more than it did computing machines. When Bond infiltrates a mine of Zorin's, he unearths a plot to detonate explosives beneath the lakes along the Hayward and San Andreas geological faults. On flooding the faults, the whole of Silicon Valley will be permanently submerged underwater. What's more, another larger bomb is placed in the mine to destroy a "geological lock" that prevents the two faults from moving at the same time.

You have to hand it to Zorin for his choice of dramatic setting. San Andreas, of course, runs through the rich and fertile state of California, and is perhaps the most famous fault line on the planet. It's where the North American Plate meets the Pacific Plate. For twenty-five million years the plates have been grinding against one another. San Andreas cuts through the very fabric of life in modern California—through roads, bridges, and communities. And the reason that twenty million people still live along the fault is that the danger zone has made California rich—from the gold rush, huge amounts of oil, and a microclimate that makes one of the richest farming regions in the whole of America. Zorin may well be unaware of the math: the next major earthquake in the fault will cost around $250 billion worth of damage. And yet California earns around $100 billion every year because of the fault. Ten thousand years since civilization started, the benefits of living along a fault line are as rich as they ever were. So thank heavens 007 put an end to Zorin's plan. For his efforts, and as a going away gift for Moore's Bond, General Gogol awards 007 the Order of Lenin.

# THE LIVING DAYLIGHTS (1987)
# MASTERS OF WAR

| *The Living Daylights* | Year: 1987 | Bond: Timothy Dalton | Director: John Glen |
|---|---|---|---|
| Producer: Albert R. Broccoli & M. G. Wilson | Screenplay: R Maibaum & M. G. Wilson | Distributor: United Artists | Release Date: June 29 |
| Running Time: 131 minutes | Budget: $40 million | Box Office: $191.2 million | Body Count: 42 |

I think Roger was fine as Bond, but the films had become too much techno-pop and had lost track of their sense of story. I mean, every film seemed to have a villain who had to rule or destroy the world. If you want to believe in the fantasy on screen, then you have to believe in the characters and use them as a stepping-stone to lead you into this fantasy world. That's a demand I made, and Albert Broccoli agreed with me.

—Timothy Dalton in "Celebrating Timothy Dalton's James Bond" by Mark Harrison, *Den of Geek* (October 29, 2012)

And I hope that you die/And your death will come soon/I'll follow your casket/On a pale afternoon/And I'll watch while you're lowered/Down to your deathbed/And I'll stand over your grave/'Til I'm sure that you're dead.

—Bob Dylan, "Masters of War," 1963

## A DARKER BOND

Having said goodbye to one Bond in Roger Moore in 1985, the search began for a new Bond. Among the actors in the frame was Northern Irish actor Sam Neill. In 1983, Neill had appeared in a television miniseries, *Reilly, Ace of Spies*, a drama about Sidney Reilly, a Russian Jew who became one of the greatest spies ever to work for the British. Also in the frame was Pierce Brosnan, another actor from Ireland. Brosnan was eventually ruled out due to his contract with *Remington Steele*, an American television series that, a little like Moore's Bond, blended the genres of romantic comedy, drama, and detective procedure.

The Bond producers then went for Timothy Dalton, a Welsh-born, classically trained Shakespearean actor. Dalton was hired in August 1986 on a salary of $5.2 million. He actually had previous discussions with Broccoli about playing Bond. But at that time, at the age of around twenty-five, the conscientious Dalton had considered himself too young to play the role, considering Bond's age to be more of the thirty-five to forty-year-old mark. Typical of a classically trained actor, Dalton, a green-eyed, dark-haired and slender six foot two inches, was keen to portray the Bond character as accurately as he could, which for Dalton meant studying extensively the original works of Fleming.

Dalton's Bond was spy-fi serious. This was a Bond cold and dark, showing little humor, but plenty of emotion. (In *The Living Daylights,* Bond's friendship with British diplomat Saunders is given touching weight, and makes us realize we've hardly seen Bond making, or losing, a male friend.) Elsewhere, Bond is a ruthless killer focused on the mission; a man with little time for fun and frolics. The interpretation of Bond came from Dalton's desire to "see a darker Bond," one which would reemerge later with Daniel Craig's incarnation. Less of a womanizer, tougher, and closer to the darker character from Fleming's fiction.

Bond screenwriter Richard Maibaum had named Sean Connery as the best Bond. But he also thought Dalton to be the best actor with whom he'd worked. Another Bond writer, Tom Mankiewicz, praised Dalton's androgynous demeanor. And Dalton's predecessor, Roger Moore, felt that Dalton was the most accomplished actor to play the role of 007, while director John Glen felt Dalton was way ahead of his time in the way he portrayed Bond.

## TAKING UP THE TUX

For many people, *The Living Daylights* is one of the best Bond films. For them, the ongoing crime of neglecting Timothy Dalton's Bond is akin to ignoring the unpopular, but secretly brilliant, student at the back of the class. The intricate plot of the movie is perhaps the franchise's closest to John Le Carré. It's a turbo-charged tale of KGB defection, the New World Order (to which we shall return in later movies), but mostly the arms trade. When you think about Daniel Craig's Bond, along with the Jason Bourne series of films, you can appreciate how this Bond movie prefigures them by fifteen years. Rather than reading Dalton as humorless, his gritty, gentlemanly Bond merely seems to possess a more refined charm than Bourne or Craig's Bond. Dalton, the antithesis of the wisecracking Moore, was true to his promise on taking up the tux. When Dalton talked about his approach to Bond as having "a sense of responsibility to the work of Ian Fleming," it highlighted how far the franchise had strayed from its spy-fi roots.

After many years of Roger Moore meaning less with Bond, comedy was hardly the first priority for the producers when *The Living Daylights* was released in 1987. Dalton's more nuanced and sophisticated interpretation arrived when 007 was in big danger at the box office. And this fresh start, for a franchise that had seen twenty-five years of gadgets, guns, and girls, and sometimes approached mere parody, was welcomed by fans and critics alike.

## *THE LIVING DAYLIGHTS*

007 is tasked with protecting a key asset, Soviet defector and KGB officer Georgi Koskov, from a lurking assassin. The assassin is a beautiful blonde cellist who has previously caught the more discerning eye of Dalton's Bond. Not only that, but Bond declines

to kill the sightly sniper. Had it been Moore's Bond, the assassin would have been won over with a swift succession of smart-ass one-liners and perkily raised eyebrows. But Dalton is different. Dalton draws on Fleming's fiction, the hidden aspects that hint at a sense of Bond's interior life; his hurricane room. Bond's new-found clipped vowels and abrupt northern English manner nonetheless suggest a depth that hides the "sharp pang of longing" he feels for the assassin cellist. When Dalton's Bond is distilled, a mini-masterwork emerges.

Bond helps Koskov defect during a symphony performance. During his debriefing, Koskov tells Bond that a new policy of assassinating defectors has been executed by new KGB head Leonid Pushkin. And yet, as Bond explores this alleged Soviet threat, a counterplot surfaces. The new narrative centers around a dodgy American arms dealer by the name of Brad Whitaker. Whitaker is a military enthusiast with a capitalist twist. A self-styled general, he has been expelled from West Point for cheating and has never served as a military officer. His corrupt trade in arms is to supply weapons to the "enemy," the Russians (who else?), and taking down payments in diamonds (this is Bond, after all). Finally, there's some irony in Whitaker's demise. Bond uses a plastic explosive to topple a bust of Wellington onto Whitaker, killing him.

## LORDS OF WAR

But what are the facts behind the spy-fi intrigue of arms trading in *The Living Daylights*? It seems that many nations, including the huge majority of the planet's large-scale arms importers, don't have the tools or perhaps the will to stop corruption in the weapons trade. Take Bond's own country of Britain, for example. A number of Britain's most profitable and potential markets for arms include Oman, Indonesia, and Saudi Arabia, all countries that have a very high risk of corruption. Transparency International (TI) is a charity set up to combat corruption and promote good practice in commerce and industry. They've been the first to report studies that act as an index to measure how, or indeed whether, governments actually fight against corruption in arms trading.

No wonder someone thought Bond should fight against the arms dealers. TI has estimated that the global cost of corruption in the weapons trade totals at least $20 billion each year. That's the same amount as the sum pledged by the G8 countries in 2009 to fight world hunger. The Director of TI's defense and security program, Mark Pyman, has stated in *The Guardian*:

> Corruption in defense is dangerous, divisive, and wasteful, and the cost is paid by citizens, soldiers, companies, and governments. Yet the majority of governments do too little to prevent it, leaving numerous opportunities to hide corruption away from public scrutiny and waste money that could be better spent.

The analyses of TI look at measures by eighty-two countries to reduce corruption risks. Such countries accounted for 94 percent of global military expenditure in 2011, equivalent to $1.6 trillion. Around 70 percent of countries leave their economy open to waste and security threats. That's because they lack the tools to stop corruption in the defense sector. Indeed, around half of those countries' defense budgets actually lack transparency entirely. And in 70 percent of countries, citizens receive no indication of how much is spent by their government on classified weapons projects.

Proper democratic accountability is rare. The TI report suggests that only 15 percent of governments have political oversight of their arms trade, which is exhaustive, accountable, and useful. In 45 percent of countries, there is little or no oversight of defense policy. And in half, there is minimal evidence of scrutiny of defense procurement. Oliver Cover, the main author of the TI study, said, "Governments should clean up this sector, and our report will give them practical solutions to achieve transparency. Doing so will save the lives of troops and citizens, and governments billions of dollars."

## A BETTER MISSION FOR BOND

What kind of role do intelligence services play in the arms trade? The truth is, rather than catching and stopping corruption as is implied in *The Living Daylights*, intelligence is actually used *in support* of arms traders. They provide their national companies with intelligence on the negotiating positions of foreign, and thus rival, manufacturers. Naturally, there's a democratic science to this. It raises crucial questions about the role of the state in the private sphere.

In 2008, the United Nations declared that $30 billion per year could end hunger on earth. That situation has changed little in the last decade. In a different report, published in 2015, the UN calculated a figure of $265 billion as the cost per year, for fifteen years, to permanently eliminate extreme poverty. This new figure would eliminate starvation and malnutrition, a somewhat broader project than simply preventing starvation, one year at a time. In 2017, the annual Pentagon base budget (which includes the war budget, the nuclear weapons in the Department of Energy, Homeland Security, and other military spending) totaled well over $1.7 trillion. This was before Congress boosted Pentagon spending by another $80 billion in the 2018 budget and passing considerable increases in further spending. Just 2 percent of US military spending could end starvation on earth. Around 16 percent of that $1.7 trillion US military spending for fifteen years could permanently end extreme poverty globally. Now there's a mission for Bond. Maybe he should have a word with Felix Leiter.

# LICENCE TO KILL (1989)
# THE CHEMISTRY OF DEATH

| Licence to Kill | Year: 1989 | Bond: Timothy Dalton | Director: John Glen |
|---|---|---|---|
| Producer: Albert R. Broccoli & M. G. Wilson | Screenplay: R. Maibaum & M. G. Wilson | Distributor: United Artists | Release Date: June 29 |
| Running Time: 131 minutes | Budget: $40 million | Box Office: $191.2 million | Body Count: 23 |

So, who has got away with murder? Not, of course, the British government. They merely covered up, as one does, the offensive corpses. Though not literally. That was done by person or persons unknown. So, who has committed murder? Not, of course, the highly respectable firm of KDH Pharmaceutical, which has enjoyed record profits this quarter, and has now licensed ZimbaMed of Harare, to continue testing Dypraxa in Africa. No, there are no murders in Africa. Only regrettable deaths. And from those deaths we derive the benefits of civilization, benefits we can afford so easily . . . because those lives were bought so cheaply.

—Jeff Caine screenplay, *The Constant Gardener* (2005)

## *LICENCE TO KILL*

Dalton's Bond was short-lived. He appeared in just two films as a dark and promising 007, *The Living Daylights* and *Licence to Kill*. Ongoing litigation over the licensing of the franchise meant that what would have been Dalton's third movie was delayed by several years. After his six-year contract expired in 1993, Dalton left the franchise in 1994. By then, Dalton's movies had not done so well at the box office. Some critics, yet again, began to say that the franchise had run its course. This was an age of movies such as *Indiana Jones* and *Lethal Weapon*. For all his Shakespearean standing and claims of a truer Bond, many thought a stage actor like Dalton was no competition as a convincing action hero in the age of Schwarzenegger and Stallone.

In some ways, Dalton's Bond now drew back further from the spectacle that is spy-fi. We quoted earlier how Dalton seemed to protest against the "fantasy on screen" that was Bond. Dalton suggested that, with the previous Bond movies, "every film seemed to have a villain who had to rule or destroy the world." Perhaps that's why some critics believe Dalton deliberately played Bond as a ruthless and sober spy, possessing little of the wit shown by Connery or Moore. As Dalton himself put it, Bond is a "man, not a superhuman; a man who is beset with moral confusions and apathies and uncertainties, and who is often very frightened and nervous and tense." Hardly Schwarzenegger and Stallone, indeed.

Dalton's portrayal of Bond was the truest and most literal reading of the role seen on screen to date. What's more, Bond now had an edge; a modicum of moral ambiguity. And so, in *Licence to Kill*, Bond goes off radar. He becomes a rogue agent. In this new movie, Bond becomes more self-absorbed, even reckless. Some scenes are brutal, by the usual Bond yardstick, with the movie as a whole suggesting a Bond prone to nervous laughter and certainly a little unhinged. Hardly surprising that, in light of *Licence to Kill*, Dalton's Bond has been called a "muscular vigilante." Dalton's films have a hard-edged reality, and unflinching violent scenes, that became the norm with Daniel Craig's Bond and the Bourne films.

## A DIRTY TRADE AND A ONE-MAN MISSION

*Licence to Kill* is the movie where Dalton's Bond goes graphic-novel-dark. It's interesting to note that, in a CBC television interview in 1963, Fleming gave his thoughts on the world of James Bond:

> Spying is in fact a dirty, dirty trade. And so is private detective work, and all that underworld of, sort of, policeman-ship is a dirty life; let's face it. And James Bond is occasionally in a dirty trade . . . [but] because perhaps the books have pace and plenty of action, and espionage is not regarded by the majority of the public as a dirty trade, they regard it rather as a sort of very romantic affair, you know. Spying has always been regarded as a very romantic one-man job, so to speak. One man against the whole police force, or an army.

And that's what we get with *Licence to Kill*. The plot becomes more of a one-man job for Bond than ever before. 007 relinquishes his licence to kill, disobeys his orders from above, and goes on a mission of revenge. And there's a dirty trade at the root of Bond's vengeance—the drugs trade. The wife of Bond's good buddy Felix Leiter is killed by drug baron Franz Sanchez. So, a beautiful (naturally) CIA pilot flies Bond to Sanchez's South American headquarters where, disguised as a hit man, Bond is hired by the most powerful drug lord in Latin America.

What follows feels very personal. It's a one-man vendetta to hunt down those who hurt his friends, then either kill them one by one, or ruin their entire lives. As Bond is on his own, we become his accomplices. We witness the killings and the dispensation of tough justice. We're at Bond's shoulder as he worms his way into the kingpin's lair. We watch as Bond sits like a patient predator until the time is ripe to make his kill. And yet, all Bond's skills are wasted on a storyline that was, by now, somewhat clichéd in cinema.

## BIG PHARMA'S DARK UNDERSIDE

By the time of *Licence to Kill*, Hollywood had been making movies about drug cartels for two whole decades. British spy author John Le Carré's 2001 novel, *The Constant Gardener* (made into a movie in 2005), goes after "Big Pharma," the term used to refer collectively to the global pharmaceutical industry. Le Carré's novel tells the tale of British diplomat Justin Quayle. Quayle's wife is an activist who's been murdered. Believing there is something behind the murder, Quayle does what Bond should have done: he tries to uncover the truth about an international conspiracy of corrupt bureaucracy and pharmaceutical money. A different kind of drug cartel, official rather than illegal.

In 2001, Le Carré wrote an article for *The Nation* entitled "In the Place of Nations." Subtitled "Big Pharma's dark underside," Le Carré claimed the pharmaceutical industry offers "the most eloquent example" of "crimes of unbridled capitalism," and was thus the perfect subject for a spy story, or indeed a spy-fi movie. What then follows from Le Carré is a moral hit list of the kind of targets a more progressive Bond might have hunted down: the scandal of spiked tobacco, designed by Western companies to cause addiction and incidentally cancer in Third World communities already ravaged with AIDS, malaria, and poverty on a scale few of us can imagine; or the scandal of the impunity with which oil companies, such as Shell, trigger vast human disasters in Nigeria, displacing tribes, polluting their homelands, and resulting in understandable uprisings, which led to kangaroo-court justice and the shameful torture and execution of very brave men by a wicked and corrupt totalitarian regime.

## SELLING SCIENCE BY THE DOLLAR, BUT BOND PULLS A ROBIN HOOD

It seems giant pharmaceutical monopolies have been donating whole biotech buildings to university departments. Big Pharma has also been endowing professorships at the universities and teaching hospitals where, coincidentally and conveniently, their products are tested and developed. Bond smells a pharmaceutical rat. He finds there has been a steady flow of alarming cases recently in which inconvenient scientific findings have been buried or rewritten. The academic authors of these inconvenient reports have been hounded off their campuses and their personal and professional reputations systematically trashed by the machinations of PR agencies in the pay of Big Pharma.

Bond then finds the US president came to power on the back of a lot of greedy people, not least Big Pharma. It poured millions into his campaign, more than twice the sum it gave the other side. Some of the godfathers and grandfathers who packaged and promoted the elected president had close connections with Big Pharma. The previous president had made the mistake of trying to resist Big Pharma's draconian Washington lobby. He even suggested, heaven forbid, the release of generic drugs to people who were dying by the million for want of them. But Big Pharma won't stand for that. They'll use the courts to entrench patent law at the price of many millions of lives lost in the Third World. Bond has more than a mere battle on his hands; for many, this is a matter of life and death.

# GOLDENEYE (1995)
## *STAR WARS* AND
## SPACE FORCES

| GoldenEye | Year: 1995 | Bond: Pierce Brosnan | Director: Martin Campbell |
|---|---|---|---|
| Producer: Albert R. Broccoli & M. G. Wilson | Screenplay: J. Caine & B. Feirstein | Distributor: United Artists | Release Date: November 13 |
| Running Time: 130 minutes | Budget: $60 million | Box Office: $355 million | Body Count: 72 |

## AND BROSNAN MAKES FIVE

Enter Brosnan as Bond, James the Fifth. We first spy him in character, clad in a black jumpsuit, running along the ridge of a dam wall. Bond is miles away from nowhere and hundreds of feet above a Soviet weapons factory. In quick succession: a swallow dive, a dropped rope, and a grappling-hook gun with laser cutter as optional extra. Courtesy, no doubt, of Q. Our man is inside. Now he emerges from the shadows, we get our first good look at the world's most famous secret agent as he slays the obligatory hench-man. "Sorry," Bond chimes, the melody of his tone faintly Irish, the face tanned and confident as it comes into frame, "forgot to knock." Mere micrometers away from bursting into a self-satisfied smile, and quipping with a quality on par with Roger Moore, Pierce Brosnan has already made his mark. (In stark contrast, our first sighting of Dalton's Bond was playing paintball in Gibraltar, and his first close-up was getting spooked by a Barbary macaque.)

And yet the pre-credit sequence of *GoldenEye* isn't done yet. Bond meets another elite agent, Sean Bean's 006, and watches (spoiler?) as his colleague takes a slug to the head. Time for a quick exit—Bond evades a militia by riding a motorcycle off a cliff. Somehow, between cliff-top and a surely fatal cliff-floor, Bond manages to lay claim to a plane! Only minutes into the movie and the spy-fi dial is cranked up to the max: smart-ass one-liners, totaled vehicles, gadgets galore, cartoon communists, and a theme song from one of rock music's finest (in this case the river-deep voice of Tina Turner).

Indeed, with *GoldenEye,* Bond is back on spy-fi form. All we need do is refer to the Bondian spy-fi check list: a state-of-the-art pre-credit sequence—check; a super-villain with a base and a psychotic lust for revenge and power—check; a memorable assistant villain—check; an attractive female civilian to act as Bond's feature-length love interest—check; Bond notching up more air miles than Led Zeppelin and trashing the local architecture in at least one famous city (also like Led Zeppelin)—check.

## WHY IT'S THE END OF HISTORY, OF COURSE

The world had changed dramatically in the six years between *Licence to Kill* and *GoldenEye.* The planet had seen a number of political revolutions. *GoldenEye* was the first Bond movie to be made after the collapse of the Soviet Union, the fall of the Berlin Wall, and the end of the Cold War. In his 1992 book, *The End of History and the Last Man,* American philosopher Francis Fukuyama said:

> What we may be witnessing is not just the end of the Cold War, or the passing of a particular period of post-war history, but the end of history as such . . . That is, the end point of mankind's ideological evolution and the universalization of Western liberal democracy as the final form of human government.

With the benefit of hindsight, Fukuyama's words later read as smug and premature. But, as the Bond producers were stuck in the early 1990s, they must surely have had some doubt about their character's relevance in the post–Cold War world. Incidentally, Fukuyama admitted in 2014 that his "end of history" theory now "obviously looks wrong." And Timothy Stanley and Alexander Lee of *The Atlantic* said in September of the same year:

> Today, it's hard to imagine Fukuyama being more wrong. History isn't over and neither liberalism nor democracy is ascendant. The comfy Western consensus he inspired is under threat in ways he never predicted. A new Cold War has broken out. China's "Marxist capitalism" suggests you can have wealth without freedom. And the advance of ISIS may herald a new, state-oriented Islamic fundamentalism.

Plenty in there for the Bond producers to pick on!

Having said all that, *GoldenEye* begins with (yes, you've guessed it!) Soviet Russians. Should moviegoers be in any doubt about the political changes since *Licence to Kill*, the movie makes the point with the subtlety of a flying mallet. The opening title sequence of *GoldenEye* features the collapse and destruction of several structures associated with the USSR, such as a red star, statues of Communist leaders (especially Joseph Stalin), and the hammer and sickle. No doubt the titles were meant to be some kind of catchup for the audience. You know the kind of thing: "Hey, folks, since you last met Bond, some serious stuff has happened; communism has fallen down and Uncle Joe's horses and men can't put it back together again." Not everyone saw the lighter side of this attempt by the franchise to do a Fukuyama, Bond style. According to producer Michael G. Wilson, some countries protested against "symbols being destroyed not by governments, but by bikini-clad women," especially certain parties in India, who threatened to boycott the film. And yet, given some of its history, a hammer and sickle being pulled down by bikini-clad women is *exactly* what one might have expected from the franchise.

## ONATOPP OF THINGS

The plot of *GoldenEye* can't leave Russia alone. It moves from Arkhangelsk to Monte Carlo, back to St. Petersburg, and finally to the Cuban jungle. St. Petersburg gets the worst of it. In this old Russian city, Bond goes joyriding in a tank! You'll remember that in this book's chapter on *Live and Let Die*, we talked about the Bondian motif of the car, and motoring about in Bond movies, and how *Diamonds Are Forever* had raised the bar on the Bond car chase. Well, *GoldenEye* goes one step further, as 007 drives a T-55, the main battle tank for armored units of the Soviet Army during the Cold War. In the T-55, Bond does a Led Zeppelin on the city, spending six expensive minutes smashing the place to pieces. When card-playing in Monte Carlo, Bond comes across a cigar-smoking murderess with the spy-fi name of Xenia Onatopp. Okay, it's not up there with Pussy Galore, Holly Goodhead, Octopussy, Plenty O'Toole, Chew Mee, or Molly Warmflash, but several times Onatopp is a threat to Bond's life. Incidentally, if you're disbelieving about some of the names in the Bond franchise, check out the book *Potty, Fartwell and Knob: From Luke Warm to Minty Badger—Extraordinary But True Names of British People*, written by Russell Ash. It includes genuine British names, such as the unfortunate "Henry Fuckingham," born in Kettleby in 1797, the incredibly Bondian "Jet Lust," born in Oxfordshire in 1840, and what one would hope is not an occupational name in origin, "Dick Wacker," born in Essex in 1781, plus many, many more. So, yes, people of the English-speaking world, some British folk really *do* have curiously unbelievable names.

You'll also recall that in this book's chapter on *From Russia with Love*, we mention Brosnan's Bond as something of a technical peak in the time-honored tradition of bumping off Bond . . . slowly. And sure enough, *GoldenEye* has its lion's share of assassination attempts. This includes not just the scene where Bond is strapped into a helicopter that is programmed to launch heat-seeking missiles at itself—surely a true classic in the category of the need-lessly elongated death sentence—but also the scene where the Bond baddie with a lair, in this case Sean Bean, tries to drop Brosnan off the Arecibo Radio Telescope dish in Puerto Rico, which doubles up as Cuba for the movie. (Bean's baddie, Alec Trevelyan, also appears to possess an evil train, which in turn possesses a special Bond-murdering carriage.)

## THE KINGPIN OF BETRAYAL

In fact, Bean's baddie is the kingpin about which the entire plot revolves. The movie traces Alec Trevelyan's trajectory from a vic-tim who's shot in the head in the film's early minutes, to becoming a bad and betraying super-villain. Once one of our men on the inside, Trevelyan has returned to torment Her Majesty's Secret Servant because as a secret Russian he secretly hated England (no doubt for being on the winning side of Fukuyama's end of history). Alec Trevelyan gets all the best lines. For example, when challenged by Bond about his motives, Trevelyan's reply taunts Bond about his easy womanizing and his unfeeling murder of many henchman over the years: "Oh, please, James! Spare me the Freud. I might as well ask you if all the vodka martinis ever silence the screams of all the men you've killed. Or if you find forgiveness in the arms of all those willing women for all the dead ones you failed to protect."

Trevelyan even conjures the image of a possible funeral for Bond: "A small memorial service, with only Moneypenny and a few tearful restaurateurs in attendance." Delicious.

When we first see Bond and Trevelyan working together back in the 1980s, they'd been scoping out a chemical weapons plant in some remote mountain range near Arkhangelsk. After infiltrating the weapons facility and planting explosives, Bond's closest com-patriot Trevelyan is captured and apparently shot as Bond makes a miraculous escape, while also managing to blow the plant to kingdom come. Now it's 1995, and Bond is on the tail of former Red Air Force jet jockey Xenia Onatopp (yes, her again). Onatopp is bad-ass. She's an assassin who uses pleasure as her ultimate weapon. And she's in the pay of a Russian crime syndicate (where's Stalin when you need him?) known as Janus. Bond is unable to stop Onatopp from stealing a powerful stealth chopper. The theft is dismissed by Bond's new female boss, MI6's new leader played by Judi Dench, a woman who taunts Bond into suggesting she is merely a bureaucratic bean counter and who, despite her acidic responses to grumbling from her disgruntled male agents, is having to learn her new role on the job, as it were. She nevertheless has to trust Bond's instincts when a nuclear accident occurs over a region in Russia that also houses the tracking station for a weapons satellite known as GoldenEye. Bond's orders are delivered by this female M in an exchange that attempts to bring 007 up to speed on contemporary sexual politics:

> **M:** You don't like me, Bond. You don't like my methods. You think I'm an accountant, a bean counter, more interested in my numbers than your instincts.
> **Bond:** The thought had occurred to me.
> **M:** Good, because I think you're a sexist, misogynist dinosaur. A relic of the Cold War, whose boyish charms though wasted on me obviously appealed to the young woman I sent out to evaluate you.

Elsewhere the movie's sexual politics are, naturally, confused. The film's heroine, Natalya, is portrayed as a firebrand. This is a woman who can look after herself. She hacks computers, she loads a semi-automatic with ease, and so on. And yet she also has an (unfunny) running joke in which she nags Bond for severely lacking in chivalry.

## *GOLDENEYE* FICTION AND FACT

Bond tracks down the stolen gunship, and all leads to a second GoldenEye system. Along the way, Bond is horrified to find the true identity of the Janus syndicate in Trevelyan. This is a Bond movie back to its spy-fi best, with a spectacular running pursuit from the stolen gunship through a Soviet army prison, to a wild car chase involving a tank, a confrontation with a Soviet missile train, and eventually a showdown with Trevelyan in the jungles of Cuba. The movie isn't perfect. For one thing, the ever-present product placement here reaches unprecedentedly hysterical levels. Bond keeps on brandishing his Omega Seamaster, for no apparent plot reason, and in the tank chase scene he drives through a clearly branded stack of mineral water. Not to mention the unscientific spectacle in another scene of a torrent of bullets deflected, with accompanying "ping" noises off a chain-link fence.

And yet *GoldenEye* scores highly in that other important category of spy-fi movies—the lavishness of the super-villain's assassination. In *Goldfinger*, Sean Connery's Bond made sure that Auric Goldfinger got sucked out of an airplane window. In *Moonraker*, Roger Moore's Bond condemned Hugo Drax to a hauntingly lonely death in outer space. But Brosnan's Bond goes one better in *GoldenEye*. Trevelyan is dropped many hundreds of feet onto concrete. And just in case that isn't fatal enough, a satellite dish the size of a basketball court lands on him. Pointy antenna-end first. And on fire. As Trevelyan screams a last desolate scream, Bond exits the scene, tethered safely to the bottom of a chopper, his female love interest at the controls. Compared to the checkered history of the sexual politics of Bond's spy-fi universe, this is progressive.

Finally, consider the scientific question of GoldenEye itself. A purely spy-fi space weapon that could destroy the Earth in a single pulse! Unbelievable? Maybe. And yet, it seems, some American presidents think such a super-weapon is easily achievable. On the evening of March 23, 1983, President Ronald Reagan addressed the American people in a televised statement. The world was a dangerous place, Reagan said. So he was seeking a technological fix in his Strategic Defense Initiative (SDI), later to be dubbed the "star wars" initiative. Who knows, maybe it was Reagan's call for space-based weapons, especially nuclear-powered X-ray lasers, which led to the plot of *GoldenEye*. The reality of Reagan's weapon never lived up to the hype. And yet that hasn't dissuaded Donald Trump from talking about his space force. Apparently, it's going to be the best ever space force, of course. Perhaps just the start of a long journey to building Trump's GoldenEye. Trump likes gold.

# TOMORROW NEVER DIES (1997)

# INFORMATION WANTS TO BE FREE

| *Tomorrow Never Dies* | Year: 1997 | Bond: Pierce Brosnan | Director: Roger Spottiswoode |
|---|---|---|---|
| Producer: Barbara Broccoli & M. G. Wilson | Screenplay: B. Feirstein | Distributor: United Artists | Release Date: December 9 |
| Running Time: 119 minutes | Budget: $110 million | Box Office: $333 million | Body Count: 54 |

**Roebuck**: And instead of decisive action, all you want to do is . . .

**M**: My goal is to prevent World War III, Admiral, and I don't think sending an armada into the recovery area is the best way to do it.

**Defence Minister**: Where exactly did this mysterious GPS signal come from?

**M**: We're still investigating.

**Roebuck**: "Investigating." With all due respect, M, sometimes I don't think you have the balls for this job!

**M**: Perhaps. But the advantage is, I don't have to think with them all the time.

—Bruce Feirstein screenplay, *Tomorrow Never Dies* (1997)

On the one hand you have—the point you're making, Woz—is that information sort of wants to be expensive because it is so valuable—the right information in the right place just changes your life. On the other hand, information almost wants to be free because the costs of getting it out is getting lower and lower all of the time. So, you have these two things fighting against each other.

—Conversation between Stewart Brand and Steve Wozniak, first hackers conference (1984)

**Elliot Carver:** Don't you realize how absurd your position is?!

**Bond**: No more absurd than starting a war for ratings!

**Carver**: Great men have always manipulated the media to save the world. Look at William Randolph Hearst, who told his photographers "You provide the pictures, I'll provide the war." I've just taken it one step further.

**Bond**: [kills a henchman sneaking up on him] Sorry, I tuned out for a moment, Elliot.

**Carver**: Touché.

—Bruce Feirstein screenplay, *Tomorrow Never Dies* (1997)

## BALANCING BOND: GETTING THE SPY-FI JUST RIGHT

When Brosnan first appeared as Bond in *GoldenEye*, the franchise got it spot on. We warm at once to James the Fifth. It helps that we see him in the post-credits sequence, driving the iconic Aston Martin DB5, as seen in *Goldfinger* and *Thunderball*. The memory of Connery's Bond immediately springs to mind. Then there's Brosnan's characterization of Bond. It's a portrayal that is certainly closer to the Fleming novels than Moore's, it's about as light and as intense at times as Connery, and certainly less dark than Dalton's Bond.

With each regeneration of Bond, some serious tweaks in character need to be made. And those changes need to work. In a 1996 issue of *Sight and Sound*, a British monthly movie magazine published by the British Film Institute, critic José Arroyo wrote about *GoldenEye*, "We want to like most movies we pay to see, but we already know the Bond formula—it has already earned our good will—so our pleasure revolves around seeing how the filmmakers execute their turn." With Brosnan, the franchise knew tweaks were needed in sympathy with changes in public attitudes. For example, Brosnan was adamant he wanted to change Bond's smoking habit. In Brosnan's words, "I don't give a damn about everyone's perception of the character: I think smoking causes cancer; therefore he doesn't smoke," although Brosnan's Bond does enjoy a brief Cuban cigar in *Die Another Day*.

A major change was Bond's attitude to women. Denounced from the get-go by a now-female M in *GoldenEye* as a "sexist, misogynist dinosaur, a relic of the Cold War," Bond must now strike a balance. Brosnan had to play a new and yet still quintessential Bond in appearance and manner. That legendary air of coolness and elegance, and an easy grace that makes Bond believable as an assassin, but with less of the playboy misogyny. It's all spy-fi fantasy, of course. In real life, a secret agent with Brosnan's sinuous athleticism, "Armani vibe," and dark matinee idol looks would hardly stay secret for very long!

## MONOMANIACAL MEDIA MOGULS

With the Cold War long gone, Bond was on the hunt for new targets, new super-villains. The franchise may have seriously glossed over the truth about the way intelligence services deal with the arms trade in *The Living Daylights*. And it may have missed its mark with *Licence to Kill*, targeting tired old drug cartels, rather than Big Pharma. But with *Tomorrow Never Dies*, the franchise never falters. The production itself may have been troubled. There were numerous script rewrites, and openly unhappy actors, with Teri Hatcher even taking her frustrations to the media, ironically. And it didn't help that hands-on producer Albert R. Broccoli had recently passed. So it's a wonder the movie is as good as it is. Sure, it fared well at the box office, though probably on the back of the success of Brosnan's inaugural *GoldenEye*. But it was the only Brosnan Bond film not to open at number one, as it was kept at number two by *Titanic*, which opened the same day.

The genius of *Tomorrow Never Dies* is in making a target out of monomaniacal media moguls. Elliot Carver is a British media baron. He's a kind of unholy black amalgam of Rupert Murdoch and Bill Gates. Played superbly by Welsh actor Jonathan Pryce, this splendid super-villain is officially an orphan, the illegitimate son of a German woman who died in childbirth. But Elliot has worked his way up to the top in the very modern media industry. He's now obsessed with his headline-screaming stature and needs only broadcasting rights in China to complete his global media empire. The Chinese are having none of it. (Maybe Carver is unaware of the fact that the Chinese have a reputation for superiority. Little wonder. Theirs is the longest continuous history of any country in the world; three thousand five hundred years of written history, to be exact. And even three thousand five hundred years ago China's civilization was old. The Neolithic age in China can be traced back to about ten thousand BC. And the earliest evidence of cultivated rice, found by the Yangtze River, is carbon-dated to around eight thousand years ago. An old Chinese missionary student once wrote that Chinese history is remote, monotonous, obscure, and worst of all, there is so much of it. The Chinese name for China is Middle Kingdom, or Middle Country; in Mandarin, Zhongguo. The name implies the country's assumed superior status. It dates from around one thousand BC, when the Chou empire, situated on the North China Plain, unaware and unconcerned with the civilizations in the West, believed their empire occupied the middle of the world. Everyone else, no doubt Carver included, was a barbarian.)

## GET CARVER

Carver has the power to reach every person on the planet through his Carver Media Group network. Except for the People's Republic of China, of course. They refuse to allow his insidious influence into their country. You can hardly blame them. Consider the Opium Wars. They were a series of conflicts in the nineteenth century that had a profound effect on British–Chinese relations for generations.

The Chinese had banned opium many times. They cited concerns about public morals. But private British traders nevertheless smuggled huge quantities of opium into China from India. It was the British way of balancing a trade deficit brought about by Britain's own addiction—to Indian tea. The Chinese protested strongly against the flouting of the ban. They even wrote to Queen Victoria, for all the good *that* did. Still, the British kept on trading. But soon came the crackdown by Lin Tse-Hsu, China's Opium Drugs Czar. He confiscated opium from the British traders and destroyed it. The British military response was severe, leading to the Nanking Treaty, which opened up a number of China's ports to foreign trade and gifted Britain Hong Kong. But the treaty didn't work. The peace broke out into a Second Opium War, in which the Chinese fared little better than the first, ending in another humiliating treaty.

Carver conjures a cacodemonic plan to thwart the Chinese. His MO? Simply start a war between the Chinese and the British. So when a British battleship is sunk in Chinese seas, and its deadly payload disappeared, Bond is sent on a dangerous mission against the clock to uncover the truth. All credit to the Chinese, too. They also suspect Carver is up to no good and dispatch their own agent Wai Lin, a member of the People's External Security Force of China, who proves an easy match for Bond. (Wai Lin is played by wonderful Malaysian actor Michelle Yeoh, star of many English, Mandarin, and Cantonese language productions, including *Crouching Tiger, Hidden Dragon* and *Star Trek: Discovery*.) Bond catches up with Wai Lin in Hong Kong. The two uncover a plot: Carver is happy to start a war, if only to feed the fire of his ego. Or, put very simply, the movie's plot portrays mere governments as goodies and private media capital as the baddie.

## WORDS ARE THE NEW WEAPONS, SATELLITES THE NEW ARTILLERY

*Tomorrow Never Dies* is one of the very few films, or plays or books for that matter, to satirize Australian-born American media mogul

Rupert Murdoch. Though a rather taboo topic in 1990s media, Murdoch had profound expansionist plans in China. A few years before the filming of *Tomorrow Never Dies,* Murdoch had acquired a Hong Kong company, Star TV, for $1 billion. He later set up network offices throughout Asia, as the Star deal enabled Murdoch's *News International* to broadcast from Hong Kong to much of Asia (including India, Japan, and China), potentially making it one of the biggest TV networks in the East. But Murdoch's plan didn't work. The Chinese administration enforced a clampdown on Murdoch's activities, which prevented his empire from reaching most of China. The movie is also very prescient about the role of media moguls in armed conflicts across the globe. In a telling exchange between Carver and Bond, Carver educates Bond about the ordnance of modern warfare and the science of capitalist media domination.

## BOND AS PROPHECY

Consider Pryce's portrayal of Elliot Carver. Carver is a media mogul clad in minimalist black suit and soul, looking like some kind of hybrid of Steve Jobs and Christof, Ed Harris's media Svengali in the 1998 movie *The Truman Show.* Carver surveys his media empire before banks of huge screens, which bring news from around the globe. (The scene is somewhat reminiscent of Bowie's character Thomas Jerome Newton in the 1976 movie, *The Man Who Fell to Earth*, where Newton watches multiple televisions at once to satisfy his thirst for information.)

And that's exactly how Rupert Murdoch himself, on the first day of the war with Iraq, watched the explosions over Baghdad—on a panel of seven television screens mounted in the wall of his Los Angeles headquarters. From his base, Murdoch called colleagues and friends, telling them of his pleasure that, after weeks of needless hand-wringing and media manipulation over the "weapons of mass destruction" alibi for war on Iraq, the battle had finally begun.

The Iraq war shone further light on the exceptional power in the hands of Murdoch, like Carver the chairman of a huge news corporation. Also like Carver, Murdoch is infamous for his willingness, rare among bosses of the biggest media companies, who usually like to wield their power in private, to personally mold the editorial policies of his newspapers. Newspapers that include over a dozen major English-language publications. Not to mention the media empire. Over the years, the editorial policies of Murdoch's news organizations have hewn very closely to Murdoch's own stridently hawkish political opinion. His was the loudest voice in the Anglophone world in the global "debate"(i.e. fait accompli) over the American-led war with Iraq. As Murdoch told *Bulletin*, an Australian magazine, in 2003:

> The greatest thing to come of this for the world economy, if you could put it that way, would be $20 a barrel for oil. That's bigger than any tax cut in any country . . . There is going to be collateral damage. And if you really want to be brutal about it, better we get it done now than spread it over months.

*Tomorrow Never Dies* is ahead of its time. Today, media empires battle with hackers and social media agencies to gain some kind of supremacy over the flow of information and the most basic functions of democracy. Malevolent media influences are still hard at work undermining democratic power and spreading fake news along the wires. Or, as Elliot Carver put it, "Call the president! Tell him if he doesn't sign the bill lowering the cable rates, we will release the video of him with the cheerleader in the Chicago motel room . . . And after he signs the bill, release the tape anyway."

# THE WORLD IS NOT ENOUGH (1999)
# BLACK GOLD
# MEANS BLOOD

| The World is Not Enough | Year: 1999 | Bond: Pierce Brosnan | Director: Michael Apted |
|---|---|---|---|
| Producer: Barbara Broccoli & M. G. Wilson | Screenplay: B. Feirstein, N. Purvis & R. Wade | Distributor: United Artists | Release Date: November 8 |
| Running Time: 128 minutes | Budget: $125 million | Box Office: $362 million | Body Count: 70 |

## BOND'S SPY-FI PUNCHES

Bond celebrated the end of the millennium with its nineteenth spy-fi movie in *The World Is Not Enough*. A comic thriller, much in the mold of the Roger Moore Bond movies, the production was filmed in many locations, which included Spain, France, the UK, Turkey, and Azerbaijan. Making $362 million at the box office, this exciting and endlessly inventive film pulled out all the spy-fi stops and Bond trademarks.

First, Bond himself. Brosnan had worked well. By now, he was aware of the need to pronounce even the most scandalous smart-ass one liners while still keeping a straight face. Then, there's the regular players. It's the last Bond movie in which Welsh actor Desmond Llewelyn plays Q. Inventor of all of Bond's gadgets and gizmos, Q had been played by Llewelyn in every Bond movie since *Russia With Love* (except *Live and Let Die*, of course, in which the producers had the crazy thought that it might be a good idea to dispense with Q as they felt the franchise needed less gimmicks). In this movie, after doing his bit showing Bond the latest tech on offer and training his eventual replacement, Llewelyn's Q sinks from sight and into the history of the Bond canon. Sadly, Llewelyn was killed in a car accident shortly after the film's premiere.

But how could Q *possibly* be replaced? The inspired answer was by introducing Monty Python's John Cleese, thereby adding another layer of the surreal to the spy-fi. ("Does this make you . . . R?" Brosnan quips, straight-faced, after Cleese showcases a BMW speedster with not only titanium armor, but also six cup holders!) Judi Dench returns as one of the strongest female characters in the whole of the Bond franchise. Within popular culture, Judi Dench is sometimes known as "The Dench." British rapper Lethal Bizzle allegedly coined the term in 2012 (although it's worth noting that one of Bizzle's 2010 songs was "Dench Stamina," albeit with a different meaning). "Dench" means "sick," or cool, or in British slang, "the dog's bollocks." Rather than playing her role in a condescending way, she's a fierce boss, an intelligence chief who's focused and sober, even when those with lesser mettle would have become drunk on the uproar of the spy-fi plot.

## TRADEMARK BOND

The trademark Bond girls are present, too. The franchise tendency has been to use two female "leads," a good girl who seems bad, and a bad girl who seems good. In this movie we have French actress Sophie Marceau playing Elektra King, the daughter of a magnate behind an oil pipeline running from the old Soviet oil fields to Europe, and American actress Denise Richards, playing Christmas Jones, a nuclear physicist whose expert knowledge can save or ruin the world.

*The World Is Not Enough* also features its fair share of chase sequences, of course. They come thick and fast. The powerboat chase on the Thames, which begins at MI6's postmodern head-quarters on the river, is reminiscent of the speedboat chase in *Live and Let Die*, with a superb series of stunts, including one boat somersaulting over another. The chase runs over dry land, back on the grand old river, then into a hot-air balloon. There is also a ski chase down a snowy slope, chased by hang-gliding, bomb-throwing assassin parasailers whose vehicles morph into snowmobiles (this is Bond, after all); a land chase in the BMW; and a sea chase, wherein Bond breaks into a submarine and later pursues a henchman by managing to pop outside the sub and then back in again.

## THAT STARRY, OIL-DRIVEN FUTURE

The super-villain in the movie makes a most dramatic entrance. The first time we see arch-terrorist Renard, played by Scottish actor Robert Carlyle, we see his oversized skull rise up from the floor in a hologram, finally made flesh when he moves into view. (M explains the contradictory enigma that is Renard; it seems a bullet to the brain is slowly but surely robbing him of his senses, and that "he'll grow stronger every day until he dies.") In one scene we see Bond walk around the hologram and appear to reach inside Renard's skull to trace the path of the bullet.

Finally, finishing up our list of the Bond spy-fi trademarks in this movie, there's the locations. At the start of the film, we glimpse Frank Gehry's new Guggenheim Museum in Bilbao, Spain, opened in late 1997 and hailed as a signal moment in architectural culture. Once more, Bond tries to position itself at the cutting edge of modern culture. Then there's London's Millennium Dome on the banks of the River Thames, which becomes a landing pad after a balloon explodes. But the plot of the movie revolves around one particular location—the oil fields of Azerbaijan.

## THAT OIL-DRIVEN PLOT

The modern world's reliance on oil is pivotal to the plot. British oil tycoon Sir Robert King is killed in a bombing at the MI6 headquarters. King's daughter, Elektra, inherits his fortune, which just happens to include billions of dollars' worth of oil deposits in the

Caspian Sea. M installs Bond as Elektra's bodyguard, as her new wealth attracts international interest. Bond flies to Azerbaijan, where Elektra is overseeing the construction of her family's oil pipeline.

Azerbaijan is an excellent and knowing choice of location. There are few places on the planet where oil has been easier to extract than the republic of Azerbaijan. Like a Bond movie, the hunt for oil has used cutting-edge science and tech throughout history, as the oil is often hard to find, and in distant places. Looking for oil is a little like Bond's detective work. But the tools that rock experts and geoscientists use to look for important clues for oil are hardly needed in Azerbaijan, as the Middle East became the key oil-producing region in the world.

## THE SCIENCE OF PETROPOLITICS

*The World Is Not Enough* continues the franchise's spy-fi obsession with cartoon Russians. Renard is not only ex-KGB turned terrorist, but the bullet in his head has also cut off his senses, so Renard can't smell, touch, or feel pain. And he assassinated Sir Robert. Those damn commies. Sorry, ex-commies. When the action switches to Istanbul, nuclear expert Christmas Jones comes to a realization. If the dastardly Renard were to carry out the deed of putting the stolen plutonium into a submarine's nuclear reactor, the resulting explosion would flatten Istanbul. The Russians' oil pipeline in the Bosphorus would be sabotaged. And Elektra's pipeline would remain uncontested.

Renard is right about "the bright, starry, oil-driven future of the West," of course. International conflict is often just a fight over oil. In our chapter on *The Man with the Golden Gun*, we talked about the movie's backdrop. It was that time in 1973 when the Saudis and Egyptians used oil as a weapon—to penalize the US and the West for helping Israel in the Yom Kippur war.

Pipeline diplomacy remains as relevant as ever. At the end of the Cold War, the US was left wondering how they might justify their military protectorates in western Europe and Japan. Saddam Hussain's 1990 invasion of Kuwait gifted them an answer: the US no longer needed to be seen protecting European and Asian allies from the Red Army. From now on, America would perpetually protect their allies from other alleged threats, such as the disruption of the oil supplies on which their economies depended. In other words, the US military is not in the Persian Gulf to protect oil bound for the US so much as to secure the oil supplies of Europe, Japan, and South Korea. And to implicitly blackmail Russia. That's because another key element of America's post-Cold War grand strategy has been the attempt to reduce the reliance of the EU on Russia, which provides around a third of Europe's natural gas. But, despite twenty years of US support for non-Russian pipeline routes, Europe is still highly reliant on Russian gas.

Nonetheless, by self-appointed policing of vital regions such as the Persian Gulf on behalf of their industrial allies, Washington hopes to keep the world safe and secure, including energy security, from the other great powers. George W. Bush expressed this idea in a 2002 address at West Point: "America has, and intends to keep, military strengths beyond challenge, thereby making the destabilizing arms races of other eras pointless and limiting rivalries to trade and other pursuits of peace." So, let's not be too hard on Renard. We could be too crude about it, but often it's possible to reduce world politics to nothing but a clash for oil. And a grasp of global petropolitics is crucial to understanding the dynamics of today's world.

America's European allies have reduced their military capabilities over the years, persuaded that Uncle Sam would always protect their energy supply if things proved tricky. So watching *The World Is Not Enough* reminds us that, though the old certainties of the Cold War may have disappeared, the world still has to remember that petropolitics still matter.

# DIE ANOTHER DAY (2002)
# SECOND SUNS AND ARTIFICIAL DAYLIGHT

| *Die Another Day* | Year: 2002 | Bond: Pierce Brosnan | Director: Lee Tamahori |
|---|---|---|---|
| **Producer:** Barbara Broccoli & M. G. Wilson | **Screenplay:** N. Purvis & R. Wade | **Distributor:** United Artists | **Release Date:** November 20 |
| **Running Time:** 133 minutes | **Budget:** $145 million | **Box Office:** $435 million | **Body Count:** 51 |

**Gustav Graves:** As you know, I try to give the planet something in return for what it's given me. Those little shards of heaven known as diamonds. Now, diamonds aren't just expensive stones. They are the stuff of dreams. And the means to make dreams real. Imagine being able to bring light and warmth to the darkest parts of the world. Imagine being able to grow crops the year round, bringing an end to hunger. Imagine a second sun, shining like a diamond in the sky. Let there be light. I give you Icarus! Icarus is unique. Its miraculous silver skin will inhale the sun's light and breathe it gently upon the earth's surface. You have no idea how much Icarus is about to change your world. And now let us brighten this night with our inner radiance.

—Neal Purvis and Robert Wade screenplay, *Die Another Day* (2002)

## DOES THIS STILL WORK?

*Die Another Day* was a turning point. It's a significant movie in the Bond franchise for a number of reasons. First and foremost, it's the movie that killed off Brosnan's Bond. After four films as Bond,

Pierce Brosnan said he'd wanted to do one final film. But although plans were in place to make a film for release in 2004, negotiations stalled, and Brosnan retired from the role in July 2004. It's also the movie that launched the careers of British actors Toby Stephens and Rosamund Pike into prominence, and made a Bond icon out of American actor Halle Berry.

*Die Another Day* was controversial on the Korean Peninsula. The North Korean government obviously took exception to the portrayal of their state as brutal and war-hungry. The South Koreans boycotted 145 theaters where the movie was released on December 31, 2002, because they were insulted by the scene in which an American officer issues orders to the South Korean army in the defense of their homeland. What's more, they weren't too keen on the lovemaking scene near a statue of the Buddha.

Brosnan's swan song was released around the buzz of Bond's fortieth anniversary. In celebration, the moviemakers added a number of homages to Bond history, ranging from downright obvious to the far too subtle. On the screamingly obvious end of the scale is Halle Berry's emergence from the ocean as an homage to Ursula Andress in *Dr. No*. There are nods to iconic gadgets too, such as the jetpack from *Thunderball* and the crocodile submarine featured in *Octopussy*. The relics are discovered by Bond as he looks through the junk in Q's lab. On the far too subtle end of the scale is not just Roger Moore's daughter, Deborah Moore, who plays a flight attendant, but also the fleeting moment when Bond picks up a field guide to the *Birds of West Indies*, which is a passing nod to ornithologist and bird expert, James Bond, whose name Ian Fleming, by his own admission, stole for his international secret agent.

## DECAY FROM SPY-FI TO FANTASY

And yet, despite an encouraging start, *Die Another Day* disintegrates. As many movie critics have suggested, whereas the first half of the movie is classic Bond, things begin to fall apart when the infamous ice palace is introduced. (There is a dominant theme of ice in the latter parts of the movie, and the franchise decked out the Royal Albert Hall in London as an ice palace for the film's world premiere on November 18, 2002.) Super-villain Gustav Graves's ice palace is in Iceland. And Graves uses it for the launch party of his Icarus satellite. The idea is that the human visitors radiating in the infrared, coupled with the natural cold of the climate, hold the palace in some kind of thermal equilibrium and makes the stay tolerable for guests. Later in the movie, Graves uses the Icarus satellite to melt the palace and do away with NSA agent Jinx (played by Halle Berry), though naturally she survives thanks to Bond.

Some critics think the post-ice palace part of the film is so inept that it took four years, the recasting of Bond, and a total transformation of tone to save the franchise. Witness the evidence. It's not just the dubious ice palace. There's also, and yet another, giant space laser. And an invisible car. And a North Korean super-villain with a face transplant who hooks his head up to a glowing dream machine at night. (It seems that, after gene therapy treatments, our Korean baddie who becomes Gustav Graves has the odd troublesome side effect of insomnia, i.e. he can't sleep at all. Graves explains this to the other Korean baddie, Zao, who appears to wander around the latter half of the film with diamonds glued to his face: "A couple of hours on the dream machine keeps me sane." And yet, all the moviegoers get to see of this incredible and much sought-after tech is a few flashing lights and the kind of mask one would otherwise see at a Mardi Gras in Rio.) It's not so much spy-fi as pure fantasy and sci-fi.

This slipping of *Die Another Day* from spy-fi into pure unbelievable fantasy is the main reason for the movie's disintegration. When we first encounter Gustav Graves (the sleepless, face-transplanted Korean super-villain played by Toby Stephens), he parachutes into Buckingham Palace, then meets Madonna, who just happens to be his fencing instructor. Then there's Q's introduction of Bond's new invisible car, the so-called Aston Martin "Vanish," which is "revealed" in an underground lair somewhere in London's South Bank. While it's true that scientists have been looking into "invisible" tech for some time, *Die Another Day* takes the tech to a level only really credible within a science fiction or fantasy narrative. To make an Aston Martin invisible to Bond's and our eyes, the scale of the object being hidden would have to be reduced, as the object has to be smaller, or similar in size, to the wavelength of light being used to view it. Today, almost two decades later, techniques for hiding objects are getting better, but the Vanish must have felt so incredibly far-future to Bond moviegoers that the story starts to lose its credibility. It's probably no coincidence that filming for *Die Another Day* began on January 11, 2002, at Pinewood Studios, by which time *Harry Potter and the Philosopher's Stone*, the highest-grossing movie of 2001, had already made its impact on world cinemas, with its magical fantasy and invisibility cloak and promise of fantastic feats to come later in the Potter franchise, such as flying cars.

## ENTROPIC BOND

The entropic tipping point is roughly when Bond starts swimming around in the ice palace. It's also about the same time when the camera's movements begin to be accompanied by inexplicable swooshing noises. It's as if Bond has left its very outer orbit of spy-fi and careered off into the realms of fantasy, through some wormhole between sub-genres. No wonder the next incarnation of Bond would rein in all the spy-fi beyond measure.

If you feel any of this is an exaggeration, consider the scene where Bond para-surfs through an iceberg tsunami. Now, the Bond franchise had entered the video game business many years earlier. But they really took off with the first-person shooter games version of *GoldenEye*, which got very positive reviews and sold over eight million copies. In *Die Another Day*, virtual gaming Bond seems to bleed out into cinema Bond. For example, there's the scene where a CGI-looking Bond surfs seamlessly and unbelievably through the arctic tsunami. (LEVEL UP!) There's also the sequence where Gustav Graves sports a patricidal electric robo-suit, just before he has a punch-up with Bond on a plummeting plane. The third piece of evidence of the gaming-cinema crossover is the ending. Moneypenny lies down, slips on a virtual reality headset, and passionately humps the air while a computer program, which she supposedly designed herself, enables her to kiss a computerized Bond.

## ICARUS: SECOND SUNS AND ARTIFICIAL DAYLIGHT

And yet, after all is said and done, the science prize for *Die Another Day* goes to the Icarus space satellite. Gustav Graves claimed that Icarus had been created to provide sunlight to areas of the world that most needed it. Those regions that lack solar radiation to nurture the development of farming in cold climates, perhaps. But moviegoers know better than to believe a Bond baddie. In truth, Graves was going to use the satellite as yet another space weapon. This time it's to detonate mines sitting along the demilitarized zone between North and South Korea. Then, with a clear path carved through the zone, the North would invade the South. Comic Cold War commie expansionism returns!

Surprisingly, though Icarus sounds like a plan only a super-villain would dream up, it turns out to be closer to scientific fact than invisible cars or surfing tsunamis. During the early 1990s, a group of science and tech guys working in Russia (uh oh) really did invent a device that redirected sunlight lost to space back down to Earth. Sitting up in space like a giant mirror, the device was created to lengthen daylight hours, provide solar energy for power, and even perhaps one day power spaceships. What's more, believe it or not, for a short time it actually worked. This real-life space satellite project was known as *Znamya*, or "Banner." It started in the late 1980s in the days of the old Soviet Union (look out, xenophobes) to test the type of tech that might lengthen the hours of the day in certain areas, with the aim of boosting productivity in farms and cities in the then USSR. *Znamya* was deployed almost a full decade before *Die Another Day*. The space satellite beamed light down from the earth's night sky. And, as predicted, the beam was indeed around two to three times as bright as the moon and two and a half miles wide. The beam passed across the Atlantic, cast a brief light over Europe, and finally fell upon Russia. Some observers on Earth said they could see a bright pulse, as if from a twinkling star, like Sirius. But astronauts in orbit reported the ease with which they could follow the path of light tracing across the atmosphere below. Not long later, the *Znamya* mirror burned up as it reentered the earth's atmosphere.

After this initial but limited try-out, Syromyatnikov spent years trying to replicate *Znamya*'s success. Sadly, Gustav Graves was not available. The project was considered by potential venture capitalists to cost too much money. What's worse, a follow-up satellite got caught on one of Mir's antennae, which ripped the delicate sail and the mission was scrapped. *Znamya* was ditched. Where's Blofeld when you need him?

# PART III: 2006—2019

# CASINO ROYALE (2006)
# TRADECRAFT SCIENCE:
# BOND VERSUS BOURNE

| *Casino Royale* | Year: 2006 | Bond: Daniel Craig | Director: Martin Campbell |
|---|---|---|---|
| Producer: Barbara Broccoli & M. G. Wilson | Screenplay: P. Haggis, N. Purvis, & R. Wade | Distributor: United Artists | Release Date: November 14 |
| Running Time: 144 minutes | Budget: $150 million | Box Office: $600 million | Body Count: 22 |

Today we are fighting Communism. Okay. If I'd been alive fifty years ago, the brand of Conservatism we have today would have been damn near called Communism and we should have been told to go and fight that. History is moving pretty quickly these days, and the heroes and villains keep on changing parts.

—Ian Fleming, *Casino Royale* (1953)

There's a Good Book about goodness and how to be good and so forth, but there's no Evil Book about how to be evil and how to be bad. The Devil had no prophets to write his Ten Commandments, and no team of authors to write his biography. His case has gone completely by default. We know nothing about him but a lot of fairy stories from our parents and schoolmasters. He has no book from which we can learn the nature of evil in all its forms, with parables about evil people, proverbs about evil people, folklore about evil people. All we have is the living example of people who are least good, or our own intuition.

—Ian Fleming, *Casino Royale* (1953)

Bond has always been something incredibly physical in my mind. Sean's movies were dynamic and balletic, almost, the way the violence and the action was done. And I thought I want to be involved with that. I want to be involved with that at every level because it's crucially important. It's what people go to the movies to watch.

—Daniel Craig, *The South Bank Show* (2008)

## DANIELCRAIGISNOTBOND.COM

Sometimes it seems no one wanted Daniel Craig to be James Bond. Almost three years after Brosnan's Bond had taken a flight of fantasy beyond the believable, the franchise had made a controversial announcement. At a press conference in the center of London, the producers declared that Daniel Craig would be the sixth actor to portray Bond in the series. Craig made a dramatic entrance as Bond. Tuxed-up and lifejacket-clad, the new 007 met the press after being shipped in on a Royal Navy speedboat. So very Bond. Craig announced that he had accepted the role mostly on the strength of the script for *Casino Royale*. Later, he said that "once I sat down and read the story, I just thought that I wanted to tell it . . . I'm a big Bond fan, and I love what he represents."

And yet, considerable controversy followed the franchise decision. Plenty of fans and critics expressed major doubt over whether the producers had chosen the right actor. And throughout filming for the first movie, internet campaigns such as danielcraigisnotbond.com showed not only their dissatisfaction, but even threatened to boycott the film in protest. The trouble revolved around the actor's look, apparently. Craig, they said, did not fit the tall, dark, handsome prototype. He simply didn't have that charismatic Bond persona, to which moviegoers were now accustomed. The most radical in this motley crew of anti-Craig militants referred to the new 007 as "James Blonde." In an almost reverse-Aryan ideology, they argued that, at five feet ten inches tall, the blond-haired, blue-eyed, and rugged Craig was far from fitting the traditional tall, dark, and suave actors of franchise past. Seemingly unaware of Fleming's original and similar dislike of Connery, the UK's *Daily Mirror* newspaper even ran a front-page news story with the headline, "The Name's Bland—James Bland."

## THE PERFECT 21ST-CENTURY BOND

*Casino Royale* got overwhelmingly positive critical praise. American movie director Steven Spielberg called Craig "the perfect 21st-century Bond." The movie earned almost $600 million worldwide, becoming the highest-grossing James Bond film to date. Many critics believed Craig to have been the first actor to truly nail Fleming's character in the book, with Craig coming closest to the author's original idea of Bond, this very long-lived male fantasy figure, than anyone since early Sean Connery. American movie critic Roger Ebert gave the film a four out of four-star rating. Ebert wrote that "Craig makes a superb Bond . . . who gives the sense of a hard man, wounded by life and his job, who nevertheless cares about people and right and wrong," and that the film "has the answers to all my complaints about the 45-year-old James Bond series," specifically "why nobody in a Bond movie ever seems to have any real emotions." Critic Paul Arendt, writing for the BBC, suggested that "Daniel Craig is not a good Bond. He's a *great* Bond. Specifically, he is 007 as conceived by Ian Fleming—a professional killing machine, a charming, cold-hearted patriot with a taste for luxury. Craig is the first actor to really nail 007's defining characteristic: he's an absolute swine." Yes, in his new Bond role, Daniel Craig really does look like he could kill you if he wanted to.

## THE SCIENCE OF THE BOND REBOOT

For many, *Casino Royale* is the best Bond movie. The franchise needed to redefine Bond for the twenty-first century. They needed to put considerable daylight between Craig's Bond and the fantastic excess of Brosnan's Bond in *Die Another Day*. So the producers decided on a total reboot. In fact, *Casino Royale* marks the first proper reboot of the franchise since 1962. If, that is, we define a reboot as a total sea-change, a complete reinvention of a series or franchise, which ignores all established continuity and backstory. It hadn't been done before on Bond. Sure, *On Her Majesty's Secret Service* may have introduced a new Bond in George Lazenby, but the series still held firm to the previous Connery films. And later, both Roger Moore and Timothy Dalton Bonds made reference to the death of Bond's wife in *On Her Majesty's Secret Service*.

Some might argue that there was a mini-reboot during the Brosnan era. The Brosnan movies had ushered in a paradigm shift in sexual politics. The Dench's "M" was one aspect of a strategy to reshape the franchise's gender politics at a time when Bond was thought of as "a sexist, misogynist dinosaur." And yet this wasn't a real reboot. M was clearly referring in her denunciation of Bond to the same sexist man that had gone before. But this too changed with *Casino Royale*. The slate of the past was wiped entirely clean. The new killer star in Daniel Craig was such an entirely new Bond that he is only just assigned his Double-O status. And the producers used the Bondian pre-title sequence of *Casino Royale* to set up Bond's backstory. Captured in film noir monochrome, the sequence shows Bond making his first two kills, the murders that make the Double-O man.

Was the franchise taking a huge risk? On the face of it, the tactic of rebooting Bond might seem a little perilous, to say the least. And yet reboots are often applied when a series is struggling at the box office, or when much time has passed since the previous movie. Brosnan's last Bond in *Die Another Day* had been his most successful, grossing $435 million worldwide at the box office. But with virtual gaming Bond bleeding out into cinema Bond, the CGI-heavy *Die Another Day* had passed the tipping point from espionage into comic book spectacle and excess. Bond was too unbelievable. (Daniel Craig added fuel to the fire on the question of "franchise past." Prior to Craig's arrival on the scene, Brosnan was often called the best Bond since Connery. Yet so incendiary was Craig's first turn as 007 that he immediately cast Brosnan as one of the weaker, lazier Bonds.)

So the Bond production team had to make amends. It's no coincidence that the Bond producers had recently acquired the rights to *Casino Royale*, the one Fleming title they'd not previously owned. Now, even though there had been awful TV (1954) and movie (1967) versions of *Casino Royale* outside the Eon series, this was their chance to reboot the series from the very start. After all, *Casino Royale* was the first Bond book, 007's origin story.

## SPY-FI REMAINS

They aimed for a caustic, haunted, intense reinvention of 007. And they hit their target. *Casino Royale* needed to be not just a great Bond movie, but a great movie, period. Reviewers lauded the reinvention of Bond. Some also lauded the movie's departure from the spy-fi tropes of previous films in the franchise, arguing that *Casino Royale* is great *because* it eschewed many of those motifs, which connected it with its spy-fi past. And yet that would be an artificial severing of present from past. *Casino Royale* may have had a new 007 in Daniel Craig. Bond may now be a muscular and rather sober-looking fellow in the Bourne mode. He may have brought a new determination to the role, with less of the suave flamboyance of earlier incarnations. And yet Daniel Craig's Bond is still an amalgam of the Bonds that went before: combining the cool steeliness of Connery with the sobriety of Dalton.

Earlier in this book, we defined the spy-fi of Bond as a genre that revolves around the adventures of a key character working as a secret agent or a spy. That much hadn't changed with *Casino Royale*. We also discussed the fact that, most of the time, Bond's adventures have centered on the intrigue of espionage between rival superpowers, especially during the Cold War. Even though the Cold War was now long dead, there was also Bond's other fixation: that spy-fi obsession with keeping a singular enemy super-villain, some diabolical mastermind such as Blofeld, from making his global mark. And in *Casino Royale* we have exactly such a super-villain. Not only is Le Chiffre a private banker who finances international terrorism, M also implies in the movie that Le Chiffre had conspired with al-Qaeda in orchestrating 9/11. Mads Mikkelsen makes his mark as the sinister Le Chiffre. The private banker comes from the subtle yet brutal school of Bond super-villains. Mikkelsen's Le Chiffre ranks alongside Mr. Big from *Live and Let Die*, the titular Dr. Julius No, or Telly Savalas's Blofeld in *On Her Majesty's Secret Service*. Just imagine if the Bond producers had plumped for a pumped-up clown of a super-villain such as Donald Pleasence's Blofeld, an easy target for Mike Myers and his Doctor Evil. Alongside the wired edginess of Craig's Bond, such a super-villain would have completely undermined any sense of genuine threat to 007 and pretty much ruined the plot.

If we run through the familiar plot elements and distinctive Bond themes, we can see that in *Casino Royale*, the franchise has either dialed up or dialed down the Bond motifs. Firstly, the dial is pretty steady on exotic international locations, the Bondian pre-title sequence, the assignment from M, and the odd "car" chase. The obligatory flirting with Moneypenny has been dialed down so much she doesn't even appear in the film, so too with the sci-fi sandbox sessions with Q. The usual meeting with allies like the CIA's Felix Leiter has been dialed up, so that this is the first Eon-produced Bond movie in which Leiter is played by a black actor. And finally, while the tongue-in-cheek humor has been dialed down a lot, Bond still jokes darkly as he is being tortured by Le Chiffre, saying defiantly, "I've got a little itch . . . down there . . . would you mind?" as Le Chiffre whips Bond's testicles.

## BOND VERSUS BOURNE

And yet it would be remiss to ignore the influence of Bourne on Bond. For decades, Bond had been the archetypal man's man. And, in that sense, he's influenced pop culture maybe more than any other figure in contemporary fiction. As we've seen, part and parcel of the Bondian way has been the portrayal of Bond as a kind of superhero, a larger-than-life fantasy figure whose actions finally pushed the envelope of plausibility way too far in *Die Another Day*. So when the franchise traded dramatic detonations and sci-fi plots for character development, they borrowed not just from the works of Ian Fleming, but also from the Jason Bourne movies, *The Bourne Identity* (2002) and *The Bourne Supremacy* (2004), both of which came before *Casino Royale*. *Casino Royale* is hugely more violent. And the graphic scenes at the film's beginning establish Craig as a tough and gritty Bond. Bond as a force of nature. As Neal Purvis put it in the James Bond special on *The South Bank Show* in 2008, "The very beginning of the film where he's just running, was just this sheer force; no gadgets, no cars, or anything. And he's not as good as the guy he's chasing. He's not superhuman, but he's still better than everybody."

Bourne is another reason for the Bond switch. It's a move away from the arch glamour of spy-fi toward a more intense personal story of revenge and redemption, set in a framework of the relevance of the super-spy in the world of modern espionage. There's little doubt that the Bourne films were partly influential on the development of Daniel Craig's Bond. It seems the world is sometimes falling over itself to find the perfect-fit actor who will be the next James Bond. But one thing's for sure, it won't be Matt Damon. This American treasure is far too famous for consideration for the British cinematic icon, especially when he's already careered across our screens as another hugely popular spy.

And yet there is a clear way in which Bourne is actually descended from Bond. The Bond movies are the prototype of the contemporary thriller genre. That genre has evolved from action-adventure films exemplified by movies involving leading male actors such as Arnold Schwarzenegger (*Commando*, *True Lies*), Mel Gibson (*Lethal Weapon*), and Bruce Willis (*Die Hard*) into overtly spy-fi films such as Tom Cruise in the *Mission: Impossible* franchise, and modern thrillers made by directors such as John McTiernan (including *The Thomas Crown Affair* starring Pierce Brosnan) and John Woo.

American screenwriter Larry Gross has argued that the Bond franchise represented "an entirely new super-kinetic cartoon-type action movie." Many of the motifs of the modern action thriller—the action and movement that is typical of Bourne, the structure built around a succession of set pieces, the dominance of technology and power, and the hero who rarely kills a villain without a throwaway quip—are all features of the Bond series.

Bond is showy, calm, and confident. Bourne, on the other hand, is brooding and bothered, yet serves the CIA despite himself, going about the globe making the world a safer place for American capitalism. Bond is focused and driven, knows what he wants, and gets it with poise and panache, while also being habitually and reluctantly protected by M. Bourne has less focus, but at least he knows his bosses are usually out to get him. But there's a danger in getting sucked into Bourne's ideology. Like Bourne, the system Bond serves has done more than its fair share of murdering, and is now also trying to pose a sense of moral superiority.

In short, we could view it like this: Bond is the tired old British imperialism of the first half of the twentieth century, decked out in liberal pomp and God-save-the-Queen gadgetry, but Bourne is latter-day American imperialism of the second half of the twentieth century: inconspicuous and conveniently amnesiac about past atrocities. This vision of Bourne comes from A. O. Scott of *The New York Times*; that Bourne is an "amnesiac assassin." Scott provides us with the wonderful description of Bourne as the supreme simulacrum of US imperialism, without knowing it: "A man who runs on pure survival instinct as he tries to figure out who is after him and why . . ." A personification of American empire, bossing the globe with a Protestant ethic while saving its spirit of capitalism, a moral austerity informing its killer instincts.

# QUANTUM OF SOLACE (2008)
# THE IRRESISTIBLE RISE
# OF CORPORATE POWER

| Quantum of Solace | Year: 2008 | Bond: Daniel Craig | Director: Marc Forster |
|---|---|---|---|
| Producer: Barbara Broccoli & M. G. Wilson | Screenplay: P. Haggis, N. Purvis & R. Wade | Distributor: United Artists | Release Date: October 29 |
| Running Time: 106 minutes | Budget: $200 million | Box Office: $586.1 million | Body Count: 31 |

## THE EVOLUTION OF THE BOND GIRL

For many, *Casino Royale* had maybe the best actor ever to play a Bond girl. With Eva Green as Vesper Lynd, some critics felt the franchise finally had an actor so good that even Craig struggled to keep up with her at times. With a franchise history rightfully criticized for sexism and misogyny, there are nonetheless some superb actors who have played alongside Bond. British actor Diana Rigg, for example, in *On Her Majesty's Secret Service*, has a far more positive, feminist edge to her as the late 1960s made way for a more politically progressive period. Rigg's character is so independent and gutsy that the self-confessed eternal bachelor in Bond not only declares his love, but even marries her. Another British actor, Honor Blackman in *Goldfinger*, managed to rise above the cartoon name of Pussy Galore in what many regard as the first modern manifestation of the Bond girl. Galore is a villainess who demonstrates her power over Bond through her cutting wit and physicality. At one point she pulls Bond's legs out from beneath him with laughing ease. Pussy rules the world, as Madonna once said.

There have, of course, been some major embarrassments along the way. Casting decisions for which the guilty agents should have either been fed to the *Thunderball* sharks, thrown into the New Orleans swamps with the *Live and Let Die* crocodiles and alligators, or plonked onto that *Goldfinger* plinth with a high-powered laser pointed at their deserving crotch. Such treatment would be adequate for recruiting Jill St. John in *Diamonds Are Forever*, Britt Ekland in *The Man with the Golden Gun*, and Barbara Bach in *The Spy Who Loved Me*. They're among the kind of character performances that feel more at home in *Austin Powers*, or the torrent of cheap and cheesy Euro-spy movies that taunted Bond's gilded heels in the 1960s.

Vesper Lynd is a far cry from that kind of Bond girl. Strong, seductive, exotic, and feminine without ever ending up as putty in Bond's dubious hands. In fact, it's Bond who ends up falling for Lynd, and it's tempting to imagine Green as the kind of progressive actor who wouldn't have taken the role had it not been for the fact that Bond had finally turned a corner on sexual politics.

## GREENE PLANET

*Quantum of Solace*, very possibly the worst movie title in the entire Bond canon, begins in typical 007 style: with a high-octane car chase in Siena in the Aston Martin DBS. Continuing the idea of Bond as a brutal force of nature, we soon see him engaged in a foot chase, reminiscent of both the Bourne movies and video games such as the *Assassin's Creed* series, a chase on a light motorbike, and a speedboat chase in Port au Prince, Haiti (scenes involving high-tech computer screens and surveillance also feel Bourne-inspired).

But Bond's ultimate spy-fi quarry in the movie is Dominic Greene, a world-renowned developer of green technology. Greene is *Quantum of Solace*'s super-villain, except for the fact that not very much about Greene is particularly super. Bond and M divine a shadowy global network of power and corruption that reaps billions. As Bond chases after the agents of an assassination attempt on M, all other roads lead to Greene, a perfect manifestation of the greed and power-obsessed late-capitalist. Greene is intent on securing a barren area of Bolivia in exchange for assisting a neofascist in staging a coup d'état, as the CIA looks the other way. No change there.

Late corporate power and the question of globalization is also a major concern of British spy author John Le Carré. In an interview with *Democracy Now* in 2010, Le Carré spoke about the real effect of globalization on countries, such as the situation implied by Greene and Bolivia in *Quantum of Solace*,

> Well, where I have seen globalization at work on the ground, it's a pretty ugly sight. It's a boardroom fantasy. What it actually means is the exploitation of very cheap labor, very often the ecological disaster that comes with it, the creation of mega-cities, the depletion of agrarian cultures and tribal cultures.

# SKYFALL (2012)
# THE SCIENCE OF CYBERTERRORISM

| Skyfall | Year: 2012 | Bond: Daniel Craig | Director: Sam Mendes |
|---|---|---|---|
| Producer: Barbara Broccoli & M. G. Wilson | Screenplay: J. Logan, N. Purvis & R. Wade | Distributor: United Artists | Release Date: October 23 |
| Running Time: 143 minutes | Budget: $150–200 million | Box Office: $1.109 billion | Body Count: 40 |

## BOYLE'S MAD LONDON PARADE

The opening of the 2012 London Olympics was a surreal, madcap, and often moving parade of British culture. The brainchild of British film director Danny Boyle, the opening ceremony attempted to define Britishness, in a potted history from a bucolic green and pleasant land, through the dark and belching satanic mills of the Industrial Revolution, to the wired-up internet age in which we live. Rather than trying to compete with the sheer scale and bling of the opening of the 2008 Beijing games, Boyle's parade used a torrent of jaw-dropping wit and invention. The ceremony told the tale of a thousand small stories, juxtaposing Britain's past and present, using a 10,000-strong cast and featuring a condensed greatest hits of British popular culture from mosh pit to posh pit: Beatles and Stones, Harry Potter and Mary Poppins, Sex Pistols and the Clash, and from Kenneth Branagh reciting Caliban's speech from *The Tempest* to the jewel in Britain's cultural crown: the NHS, a national health care comprehensive, universal, and free at the point of delivery.

To cap it all came James Bond. It's typical of irreverent British wit that the usually overblown affair of the arrival of the head of state should be subverted in the way that it was. The Queen played a leading role in Boyle's £27 million production. She appeared in a prerecorded film, along with Daniel Craig as Bond, in which the pair appeared to skydive down into London's Olympic stadium (incidentally, the professional stuntman and base jumper who played the Queen in the skydive went by the surname of Connery, but not *that* Connery). In the movie, global audiences watched as Craig's Bond walks into Buckingham Palace surrounded by corgis to pick up the Queen, who's seated at her writing desk and greets 007 in super-villain style with the words "Good evening, Mr. Bond." It's at that point that the pair apparently board a helicopter that flies across London to the stadium, parachuting to the ground, when only at the last moment, and in real time, does the Queen appear in the arena to take her seat to satisfied applause.

## ICONIC TELEVISION

Such quirky and esoteric wit may or may not have translated to the expected global audience of around one billion people. Not a bad start to a promo campaign for the franchise's new movie *Skyfall*, to be launched a mere three months later, and four years after *Quantum of Solace*. Across the world, Boyle's mad parade provoked plaudits, excitement, and no little amount of bafflement. In America, *The Washington Post* seemed especially galvanized by the Queen's appearance. "So, we're all watching this movie at Olympic Stadium in which James Bond (Daniel Craig) walks into the Royal Palace," wrote journalist Mike Wise. "He's followed by two mutts and suddenly walks in to see, yes, Queen Elizabeth. It's her first role. Ever!" The *Post*'s verdict? "It's corny, cheesy, altogether over the top. And it works! [...] This is awesome." (This last sentence might even sum up the entire late-Bond franchise before Daniel Craig.)

The London games had a huge impact beyond Bond itself. According to a YouGov poll for TalkTalk in the month before the release of *Skyfall*, viewers identified the Queen meeting James Bond as one of the iconic moments that changed TV forever. "What makes this television moment so iconic is the juxtaposition of the Queen and James Bond together," said Professor Jonathan Brignell. "Viewers gazed in wonder and puzzlement, pride and laughter: is this really the Queen? The moment beautifully broke the boundaries between fact and fiction, and in doing so made its mark worldwide."

## *SKYFALL'S* SPY-FI DIALS

Bond was back with a spy-fi bang. *Skyfall* premiered in London at the Royal Albert Hall on October 23, 2012, in a release timed to coincide with the fiftieth anniversary of the franchise. Very well received by the critics, *Skyfall* was lauded for its screenplay, its acting (especially Craig, Javier Bardem, and Dench), and the direction of Sam Mendes, along with the movie's cinematography, score, and action sequences. *Skyfall* was the fourteenth film to gross over $1 billion globally, and the first Bond film to make that kind of mint. Not only that, but *Skyfall* became the seventh-highest-grossing movie at that time, the highest-grossing film in Britain, the highest-grossing film in the franchise, the highest-grossing movie globally for MGM and Sony Pictures, and the second highest-grossing movie of 2012. *Skyfall* won many honors, including two Academy Awards, two Grammys, and two BAFTAs.

Not only was Bond back in a big way, but the spy-fi features of the new film were now dialed up. Having been dialed down for the reboot of the series, the classic Bond themes were reinvented for Craig's Bond. The pre-title sequence is a high-octane chase in which the Land Rover driver is none other than a black Moneypenny. The dial for exotic international locations reads

Macau, Shanghai, Istanbul, and the Scottish Highlands. And the sci-fi sandbox sessions with Q have taken an altogether new twist. Rather than portraying Q as an antiquated old boffin, Bond's new Quartermaster is a boffin of the millennial kind: all anorak, heavy rimmed specs, and computer algorithms. Q displays an attractive strain of the trait the British prize above all others: modesty. A lovely exchange between Bond and the new Q occurs when the movie revisits the sci-fi sandbox, though admittedly in more muted style:

> **Q:** Walther PPK/S 9mm short. There's a micro-dermal sensor in the grip. It's been coded to your palm print so only you can fire it. Less of a random killing machine, more of a personal statement.
> **Bond:** And this?
> **Q:** Standard issue radio transmitter. Activate it and it broadcasts your location. Distress signal. And that's it.
> **Bond:** A gun . . . and a radio. Not exactly Christmas, is it?
> **Q:** Were you expecting an exploding pen? We don't really go in for that anymore. Good luck out there in the field. And please return the equipment in one piece.
> **Bond:** [sarcastically] Brave new world.

In this scene, both Q and Bond are sitting and looking at J. M. W. Turner's famous 1839 painting, *The Fighting Temeraire*, a portrait of the soon-to-be-scrapped battered British warship. Q's insightful, poetic reading of the picture implies not just 007's naval heritage, but also that Bond himself is a soon-to-be-scrapped battered British icon. The poignant moment is soon undercut by the taciturn Bond, who sees little poetry or metaphor at work, instead saying the painting was of "a bloody big ship."

## *SKYFALL'S* SILVA

When it comes to the super-villain, the *Skyfall* spy-fi dial is turned up to the max. Raoul Silva is a former MI6 agent turned to cyberterrorism who targets the agency as part of his masterplan for revenge against M, for whom he holds a homicidal grudge. Playing this razor-sharp foil to Bond is Spanish actor Javier Bardem. Bardem has long been known as a connoisseur of the scary haircut (witness *No Country for Old Men*), to which he here adds Halloween teeth and a metrosexual menace. According to director Mendes, most of Silva's super-villain weirdness came from the actor himself, who wanted to expand the character into a real spy-fi nemesis, an aim he carries off with the very best of Bond baddie aplomb. In one scene, Silva caresses Bond's scarred body with visible relish, so Silva can perhaps be seen as a dark, dramatic, and sexually deviant nemesis who serves as a distorted reflection of 007, with Bond himself implying that he's no stranger to homosexual experience.

## IT'S IN THE SHADOWS

Wounded by the shooting and his apparent betrayal by M, *Skyfall* is a world in which Bond not only appears to have died, but also in the process becomes a virtual dead man walking. Stripped of mission and allegiance, Bond is soon back in action when news of a direct attack by Silva on MI6 headquarters reaches him. Bond returns to Britain to help protect not just his country, but also the one human to whom he feels any real personal loyalty—M.

The other major theme of *Skyfall* is the hidden enemy of cyberterrorism. We spoke earlier of Sam Mendes's use of Turner's art in the National Gallery scene with Bond and Q. This link between Bond and British art is returned to when Lord Tennyson's poem "Ulysses" is referenced by M in her wonderful address to the Parliamentary Select Committee. Musing on a discontented hero returning home to his kingdom after years of far-flung travels, Tennyson's words are beautifully recited by The Dench as we see

Bond sprint furiously across London trying to save M from Silva's latest stab at assassination.

In M's speech to the Committee she says she is frightened, as the enemies of democracy "are no longer known to us . . . our world is not more transparent now, it's more opaque . . . It's in the shadows." How profound M's words turned out to be. Less than a year after *Skyfall* hit the big screen, a British political consulting firm was set up, maintaining offices in London, New York, and Washington. The firm combined data mining, data brokerage, and data analysis with strategic communication during electoral processes. The name of the company was Cambridge Analytica (CA).

In 2016, CA worked for Donald Trump's presidential campaign as well as for Leave.EU (one of the agencies campaigning in Britain's referendum on European Union membership). CA's role in these campaigns has been controversial, to say the least. And their practice is still the subject of criminal investigations in both the UK and the US. It wasn't until March 2018 that multiple media outlets broke news of CA's shadowy business practices. One of the more prominent investigative journalists working on the CA story was British writer Carole Cadwalladr, who later won the Orwell Prize for political journalism in June 2018.

## AS THIS DARKNESS FALLS

In April 2019 Cadwalladr gave a TED talk about the links between CA, Facebook, the "right-wing fake news ecosystem," and the threat to democracy. Parts of Cadwalladr's speech are a fascinating accompaniment to M's words about those working in the shadows. As Cadwalladr put it in her TED talk:

> I don't have to tell you that hate and fear are being sown online all across the world. And we know that there is this dark undertow which is connecting us all globally. And it is flowing via the technology platforms. But we only see a tiny

amount of what's going on, on the surface. And I only found out anything about this dark underbelly because I started looking into a company called Cambridge Analytica. And I spent months tracking down an ex-employee, Christopher Wylie. And he told me how this company, that worked for both Trump and Brexit, had profiled people politically, in order to understand their individual fears to better target them with Facebook ads. And it did this by illicitly harvesting the profiles of 87 million people from Facebook.

Finally, Cadwalladr addresses the kind of cyber-forces that have preoccupied Bond in *Quantum of Solace, Skyfall,* and, as we shall soon see, *Spectre*: "We are what happens to a Western democracy when a hundred years of electoral laws are disrupted by technology."

# SPECTRE (2015)

# NEW WORLD ORDER OF SPYING: A MORE FASCIST FUTURE?

| Spectre | Year: 2015 | Bond: Daniel Craig | Director: Sam Mendes |
|---|---|---|---|
| Producer: Barbara Broccoli & M. G. Wilson | Screenplay: J. Logan, N. Purvis, J. Butterworth & R. Wade | Distributor: United Artists | Release Date: October 26 |
| Running Time: 148 minutes | Budget: $230–300 million | Box Office: $880.7 million | Body Count: 235 |

## DAY OF THE DEAD

And finally, to *Spectre*. Director Sam Mendes's second Bond adventure may not be on par with its franchise-beating predecessor. And yet it was still on target in delivering a more classic set of Bondian and spy-fi themes. Themes which its worldwide audience had come to expect from the seemingly indestructible franchise: Earth-encircling locations in Mexico, London, Rome, and Tangier; spectacular stunts; laugh-out-loud smart-ass one liners (especially those delivered by Ben Whishaw's scene-stealing Q); and a plot of impossible intrigue, which at first may seem too sci-fi and yet is actually alarmingly accurate about our brave new world of surveillance technology—intrinsically sinister and infinitely corruptible. After all, in the words of Lord Acton, a British historian of the late nineteenth and early twentieth centuries: power corrupts, and absolute power corrupts absolutely.

Bond also squares up to a superb super-villain in Christoph Waltz's Franz Oberhauser. Oberhauser, a nominal relative of a character from the original Fleming story of *Octopussy*, is introduced as a shadowy silhouette from his and Bond's shared childhood, but later becomes Blofeld—the mysterious mastermind behind SPECTRE. With Waltz's Teutonic vowels and camp but menacing manner, Oberhauser is a text-book Bond baddie, one of many nostalgic elements that make *Spectre* so spy-fi. There is a nod to "franchise past" in a scene where Oberhauser outlines the plot to his nemesis just before trying to bump off Bond with a byzantine array of robotic drills and inexplicable restraining straps.

*Spectre* opens with a long and lavish tracking shot, which appears to last a full four minutes before the first cut. The lingering shot leads us through Mexico City's Day of the Dead parade, in the tradition of *Live and Let Die*'s voodoo festivities and the Carnival Bond theme, which we talked about way back in our chapter on *Thunderball*. We are taken through the throng of the carnival itself, as the camera zooms in on a villain then veers over to a spooky couple, where we spy a dem-bones-dem-bones-Halloween-trick-or-treating-skeletal-suited Bond, whom we follow through a throbbing hotel lobby, into the ornate art nouveau-style elevator of Gran Hotel Ciudad (which also featured in *Licence to Kill*), through a window, and out on to the city's myriad rooftops before finally cutting as a gun sights its target. It's one of the best ever Bond pre-title sequences and acts as a mouth-watering curtain-raiser to a prologue that boasts jaw-dropping scenes, collapsing buildings, and loop-de-looping choppers. As *Spectre*'s Dutch-Swedish cinematographer Hoyte van Hoytema says, "It was a very visceral way for the audience to be sucked into the film."

All the initial exotic weirdness of the movie comes crashing back down to Earth in London. MI6 is being forcibly merged into a new world order of global snooping, spearheaded by the other great villain of the film—Andrew Scott's utterly unlovable Max

Denbigh, aka "C." Later, Oberhauser reveals to Bond that SPECTRE has secretly bankrolled the new Joint Intelligence Service, while also staging terrorist attacks around the world, creating a need for the Nine Eyes program. In return, C will give SPECTRE unlimited access to intelligence gathered by Nine Eyes, allowing them to anticipate and counteract investigations into their operations. It's a more fascist future.

## THE DARKER WORLD OF CRAIG'S BOND

With the advent of Craig's Bond, the world had gotten darker. The focus now became another time-honored aspect of sci-fi: the double-edged sword of technology in the shape of computer tech and surveillance. From the early 1980s on, sci-fi developed a new obsession known as cyberpunk. The climate of incorporation, with the economic power of large multinational companies wedded to high tech, led to the creation of a new kind of science fiction, so thoroughly rooted by a technophilic love of gadgetry that it in turn came to influence a whole new worldview. From its emphasis on control (of individuals, societies, and governments) it took "cyber," meaning "to steer" in Greek. From its complete rejection of that control, of that authority, it borrowed the phrasing of a similarly nihilistic cultural movement of the late 1970s—punk. Together they fuse as a portmanteau, their essential dichotomy giving a hint of the energy that resides within the movement. Cyberpunk, the sci-fi incarnation of aggressive techno-capitalism, was born. In the sub-genre, the near-future world is surveyed and controlled by computer tech. And rebellion against that control is led by lone characters, often portrayed as alienated and marginalized, just like Bond in *Spectre*.

## THE REALITY BEHIND *SPECTRE*

*Spectre* recognizes that the aggressive techno-capitalism of the 1980s has gotten far worse. The new concern is that countries in the West, and particularly Britain and America, are increasingly looking like "managed" democracies, little different from the authoritarian capitalism of China, but paid for by billionaires, using military-style tech, and unknowingly enabled by the rest of us.

The real story behind *Spectre* extends to key players in Silicon Valley and the billionaire super class. It envelops complicit governments, the military, and spy agencies intent on exploiting the internet and social media for surveillance, control, and manipulation of societies and its citizens. In 2002, the BBC aired a documentary called *The Century of the Self*. Directed by Adam Curtis, the documentary detailed how managed democracy could be achieved through propaganda and the manipulation of populations using mass media. The American military and the CIA had used psychological operations, such as emotive mass propaganda, in various theaters of war, from the conflicts in Korea and Vietnam, to Central America during the Cold War. The difference today, as the Bond franchise knows in *Spectre*, is that the internet is the new playground for twenty-first-century psychological operations.

The technological fairy tale had begun well. Once upon a time, Silicon Valley had made its play as the shiny, happy face of global capitalism. We were sold the promise of new tech, which would wire up our world into a global village. There were even whispers of the utopian idea that the internet might possibly liberate societies. And yet the double-edged sword of tech cuts two ways: soon, Big Data met Big Brother. The happy marriage of huge reservoirs of data and the politics of neoliberalism enabled the metrics of the market to be extended to every sphere of society. Big tech failed to see the contradiction in hooking up to a libertarian philosophy while at the same time being integrated with the national security state. In fact, Silicon Valley became part of the military-industrial complex, flogging its technologies to military, intelligence, and law enforcement agencies. Working for the clampdown.

Take Google, for instance. First, they brokered a deal with the National Geospatial-Intelligence Agency allowing the NGA to use Google Earth Builder. The mapping tech of Google is used for geospatial intel purposes, like supporting troops in Iraq. Google and the NGA have also bought the Bondian-sounding GeoEye-1, the planet's highest-resolution satellite. Second, Google has an open-door policy in recruiting managers with backgrounds in military and intelligence work. Google is also happy to be partners with defense contractors, such as Northrop Grumman and Lockheed Martin, focusing on robotics and drones. Likewise, Amazon created a $600 million cloud computing system for the CIA that services all seventeen American intel agencies.

## BLOFELDIAN BILLIONAIRES

Consider Silicon Valley tech billionaire, Peter Thiel. Thiel, who cofounded PayPal, is a Trump donor and backer, and also cofounder of the data-mining company Palantir. A leader in the field of mining huge data sets, Palantir's customers include the NSA (National Security Agency), the FBI, and the CIA. The data sets have been used by the marines in Afghanistan and in the Mexican drug war. Palantir's clients include JPMorgan and News Corp, Bank of America and Big Pharma.

Thiel has cultivated something of a libertarian and Blofeldian persona. In a 2009 essay called "The Education of a Libertarian," Thiel traces the increasing incompatibility of capitalism and democracy over the best part of the last hundred years. Thiel says that capitalism is "not popular with the crowd," and derides the "unthinking demos" that have made annoying and inconvenient demands for social democratic concessions, thereby constraining capitalism. Thiel's venture-capital firm Founders Fund runs an online manifesto that starts, "We wanted flying cars, instead we got 140 characters," reflecting his pessimism on the demise of technological utopianism. Like excerpts from a Bond screenplay,

Thiel obsessively complains of "economic stagnation," suggesting that without a new tech revolution, there could be a "worldwide conflagration." He is no doubt contemptuous of a 2016 Harvard study, which found not only that a majority of American millennials reject capitalism but that a third even support socialism.

## SPOOKS WORKING FOR THE CLAMPDOWN

And finally, to the real role of spies and their agencies in any future clampdown like the one suggested in *Spectre*. The so-called "war on terror" allowed the state to amass authoritarian control over citizens into something resembling an Orwellian Big Brother state. The militarization of homeland security is now standard. Predator drones prowl the Canadian and Mexican borders through real-time surveillance. The Department of Homeland Security amasses visual data on US citizens. The state has deliberately conflated the "war on terror" with domestic political activism, so terrorism can now include any enemy of the state and/or corporate interests. People like Edward Snowden and Julian Assange can be viewed as political prisoners or exiles.

Intel agencies are now mass-funding research into the use of the internet for twenty-first century psych-ops, studying social links, the propagation of messages, and the influencing of behavior. Both the NSA and its British counterpart GCHQ (Government Communications Headquarters) have been spying on social media, with a view to propagandizing, mass messaging, and pushing narratives. GCHQ, meanwhile, says its intention is to seed state propaganda across the net. Agencies are capable of scanning billions of online sources and monitoring the worldwide web to predict, and no doubt constrain, future events. Intelligence agencies monitor sites including YouTube and Facebook, attempting to control, infiltrate, manipulate, and warp online discourse and even carry out false flag operations in order to discredit targets. As Thiel has put it, "The fate of our world may depend on the effort of a single person who builds or propagates the machinery of freedom that makes the world safe for capitalism."

There appears to be a growing consensus amongst Blofeldian billionaire elites. The status quo is unsustainable, they say. The future will not only be unstable but could also see fatal systemic collapse and "civilizational crisis." They mean revolution, dear reader. In the meantime, this status quo is liable to make worse the poverty, climate change, and social divisions already present. The elites feel authoritarian clampdown is needed in the future face of increasing rioting, disruption, and civil unrest. The theater of global politics has embraced the virtual world of the internet. As American media theorist Marshall McLuhan once said, World War Three will be a guerrilla information war with no distinction between military and civilian participation. Plenty to be picked off those political bones for the Bond of the future.

# INDEX